The publisher and the University of California Press Foundation gratefully acknowledge the generous support of the Simpson Imprint in Humanities.

Meat Planet

CALIFORNIA STUDIES IN FOOD AND CULTURE

Darra Goldstein, Editor

Meat Planet

ARTIFICIAL FLESH AND
THE FUTURE OF FOOD

Benjamin Aldes Wurgaft

UNIVERSITY OF CALIFORNIA PRESS

University of California Press, one of the most distinguished university presses in the United States, enriches lives around the world by advancing scholarship in the humanities, social sciences, and natural sciences. Its activities are supported by the UC Press Foundation and by philanthropic contributions from individuals and institutions. For more information, visit www.ucpress.edu.

University of California Press
Oakland, California

Library of Congress Cataloging-in-Publication Data

Names: Wurgaft, Benjamin Aldes, author.
Title: Meat planet : artificial flesh and the future of food / Benjamin Aldes Wurgaft.
Description: Oakland, California : University of California Press, [2019] | Includes bibliographical references and index. |
Identifiers: LCCN 2019002873 (print) | LCCN 2019004418 (ebook) | ISBN 9780520968264 (ebook and ePDF) | ISBN 9780520295537 (cloth : alk. paper)
Subjects: LCSH: Meat substitutes. | Artificial foods. | Meat industry and trade—Moral and ethical aspects.
Classification: LCC TP447.M4 (ebook) | LCC TP447.M4 W87 2019 (print) | DDC 664/.929—dc23
LC record available at https://lccn.loc.gov/2019002873

Manufactured in the United States of America

26 25 24 23 22 21 20 19
10 9 8 7 6 5 4 3 2 1

For Shannon

CONTENTS

ACKNOWLEDGMENTS

Behold my debts, according to the flesh.

This book would have been impossible to research without the generosity and kindness of many people in the cultured meat world. I want to give an especially hearty thank-you to Isha Datar and Erin Kim of New Harvest, as well as to Kate Krueger, Marie Gibbons, Jess Krieger, Natalie Rubio, and Andrew Stout, and last but definitely not least to conference production expert Morgan Catalina. Mark Post welcomed me to his laboratory in Maastricht and shared his thoughts with a rare candor; my thanks to him and to his whole family, and thanks to everyone at the Post Lab. Thanks to Hemai Parthasarathy and Lindy Fishburne at Breakout Labs for conversations on the future of technology; to Oron Catts for conversation, provocation, and good jokes about biotechnology; and to Ingrid Newkirk of PETA. Ryan Pandya and Perumal Gandhi let me pal around and ask odd questions as they cultured milk proteins from Cork, Ireland, to Berkeley, California. Thanks go out as well to Bruce Friedrich of the Good Food Institute, Paul Shapiro, and Jacy Reese for illuminating conversations and exchanges. Nils Gilman was my first guide to the world of futures work and consulting, and my thoughts on futurism reflect nearly a decade of our conversations; I remain in his debt. Scott Smith and Ramez Naam also shared their thoughts on futures work and its affinities for other practices, from marketing to psychotherapy to science fiction writing. Mary Catherine O'Connor and Amy Westervelt helped me link cultured meat to the wild world of climate journalism. Michael Rudnicki corrected my errors regarding stem cell function. Cor van der Weele shared conversations about bioethics and the philosophy of biology. There are many other researchers and activists whom I'd like to recognize here, but who will remain anonymous by their request—I am grateful for their help.

Not every writer gets to play with fake chicken nuggets in the form of a dodo bird in the course of research, or hear speculations on the relationship between cultured meat and transitions from modernist to postmodernist aesthetics; many thanks to Koert van Mensvoort and Hendrik-Jan Grievink of the Next Nature Network for a terrific visit in Amsterdam.

Hail to my interlocutors at the Institute for the Future: Rebecca Chesney, Sarah Smith, Miriam Lueck Avery, Lyn Jeffery, and Max Elder. Thank you for letting me observe the practice of futures consulting in real time. I was lucky to be a frequent visitor to London in the course of my research, and I want to thank my UK contacts in cultured meat work, Marianne Ellis and Illtud Dunsford. I can't thank Neil Stephens enough; my work builds on the foundation he established through years of interviews with cultured meat's early researchers, and I benefited enormously from our conversations. This book would be a lesser thing without him. Alexandra Sexton taught me much about experiential encounters with meat surrogates as our research pathways crossed and re-crossed. David Benqué illuminated the intersections between cultured meat and the world of design. Back in the States, Christina Agapakis shared her early doubts about cultured meat and her sense of humor about the claims of biotechnologists such as herself. I thank Amy Harmon, Tom Levenson, and Nicola Twilley for conversations about the ins and outs of contemporary science journalism; and Warren Belasco, Nadia Berenstein, and Rachel Laudan for their reflections on food history, which intertwine to form a ribbon that runs through this book. Cassie Fennell coached me in the ways of ethnographic fieldwork and taught me not to let any opportunity slip by.

Ethnography is better on a good night's sleep. Jennifer Schaffner let me use her guest room in Potrero Hill, San Francisco; Jeremiah Dittmar lent me his inflatable mattress in Hackney, London; and Jordan Stein let me sleep on his comfortable couch in Hell's Kitchen, New York. Thanks, guys!

I caught my first foggy glimpse of this book's topic as an Andrew W. Mellon Interdisciplinary Postdoctoral Fellow at the New School for Social Research, where I lucked into an office next to Nicolas Langlitz, who introduced me to the anthropology of science. My thanks to him, and to my colleagues at the New School, for encouraging my research into fields beyond European intellectual history. A grant from the National Science Foundation (grant no. 1331003: Tissue Engineering and Sustainable Protein Development) would eventually allow me to spend two years as a postdoctoral fellow at the Massachusetts Institute of Technology (MIT), pursuing additional training in the history and anthropology of science. I am deeply grateful to Stefan Helmreich and Heather Paxson; Stefan sponsored my presence in MIT's anthropology department and advised my research, and he and Heather have taught me so much as friends and interlocutors. Irene Hartford, Barbara Keller, and Amberly Steward showed me how to navigate MIT. Maria Vidart-Delgado joined me to praise and bemoan the postdoctoral condition. Anya Zilberstein and David Singerman taught me about environmental history and the history of science. I thank Mike Fischer, Jean Jackson, Erica James, Graham Jones, Amy Moran-Thomas, Harriet Ritvo, and Chris Walley for friendly conversations and expert feedback during my MIT sojourn. I also thank Hannah Landecker at the University of California (UC), Los Angeles, for conversations and guidance about the history of tissue culture and for telling me about *The Space Merchants;* and James Clifford, at UC Santa Cruz, for chatting with me about the disciplinary links and caesurae between history and anthropology.

Meat Planet's somewhat experimental form owes much to the writerly company I kept during my years of research. I want to thank Alice Boone for conversations that inspired *Meat Planet*'s methodology, as well as my friends in and around Beta Level/The Errata Salon in Los Angeles and at the *Los Angeles Review of Books:* Nicole Antebi, Amina Cain, Jason Brown, Colin Dickey, Boris Dralyuk, David Eng, Ariana Kelly, Evan Kindley, Sarah Mesle, Heather Parlato, and Amarnath Ravva. The Prelinger Library and Archive in San Francisco served me as a major creative inspiration, a place to regain my bearings and find a new path when the challenges of research and writing seemed too great; my sincere thanks to Megan and Rick Prelinger. That I was able to write this book at all says much about the skill and care of my psycho-analyst, Michael Zimmerman.

My parents, Merry (Corky) White and Lewis Wurgaft, are this book's secret heroes, the ones who held me up through several apparent collapses of the world. They asked for no thanks except that I continue writing. They have dedicated much to me, so consider this a *sotto voce* second dedication of this book, to them. I also thank Gus Rancatore and Carole Colsell for their kindness during the difficult months of the manuscript's preparation.

And I am grateful to everyone who read the book in draft. Lewis Wurgaft read every chapter as I wrote it and told me exactly what he thought. All my best lines result from the push and pull of our editorial conversations. I was lucky to have his guidance. Merry White read my drafts with an anthropologist's eye and offered expert advice. So did Stefan Helmreich, who made sure that my book would be "peopled" with people as well as with their ideas. I thank Kate Marshall, my wonderful editor at UC Press, for guiding me through the shoals of the publishing process and ensuring that this book speaks to as many audiences as possible. Thanks, too, to Bradley Depew, Enrique Ochoa-Kaup, Kate Hoffman, Richard Earles, Alex Dahne, and the whole team at UC Press. I was also very fortunate that Darra Goldstein saw fit to include *Meat Planet* in her food studies series; Darra was the editor who first published my essays on food, in the early years of *Gastronomica: A Journal of Food and Culture,* and I owe her much. Thanks go out to Mike Fortun, Rachel Laudan, Tom Levenson, and Sophia Roosth for providing thoughtful, supportive, and constructive readings of the manuscript for UC Press.

Very special editorial thanks are also due to Kat Eschner, whose perceptive comments and stylistic advice have brought out the best parts of the book and made editing a joy—and to Rebecca Ariel Porte, whose careful reading pointed me toward literary *poiesis* beyond both *bios* and *zoe,* and who refused to let me take myself too seriously. Josh Berson was finishing his own book about meat as I was writing mine, and my book reflects our running conversations and readings of one another's drafts, as well as my intellectual debts to his understanding of human subsistence niches past, present, and future. Alice Boone showed me the glitches and errors of my ways, and helped me to see how each *culpa* could become properly *felix* as I came to see my argument and conclusions in a new light. My thanks for reading and commenting also go out to Elan Abrell, Christina Agapakis, Nick Barr, Warren Belasco, Shamma Boyarin, Isha Datar, Nathaniel Deutsch (who helped me name the book during a hike in Santa Cruz), Jeremiah Dittmar, Aryé Elfenbein, Nils Gilman, Erin Kim, Kate Krueger, Edward Melillo, Ben Oppenheim, Alyssa Pelish, Justin Pickard, Nick Seaver, Alexandra Sexton, Sarah Stoller, and Shannon Supple.

I gave many talks, both academic and popular, as I tried to draw *Meat Planet*'s substance into the world. My thanks to my hosts and listeners at many institutions, including the University of Pennsylvania's Wolf Center for the Humanities (and to Jim English and Emily Wilson there); to the Oxford Symposium on Food and Cookery (and to Ursula Heinzelmann and Bee Wilson there); Barnard College; UC Santa Cruz; New York University; the Huntington Library; MIT; UC Berkeley; the Errata Salon; the Lost Marbles Salon; the Institute for the Future; Forum for the Future; and Continuum Innovation.

My thanks to Aimee Francaes and Jesse Hassinger of Belly of the Beast, in Northampton, Massachusetts, who made sure that at least some of my meat consumption was local, sustainable, and delicious.

This book is dedicated to Shannon K. Supple, whose care and support have carried me through so much, and whose curiosity and sense of wonder inspire me.

Benjamin Aldes Wurgaft
Los Angeles, California
Oakland, California
Cambridge, Massachusetts

An expanded version of chapter 11, Copy, will appear as "On Mimesis and Meat Science" in the journal *Osiris*.

An expanded version of chapter 12, Philosophers, appeared as "Biotech Cockaigne of the Vegan Hopeful," in the *Hedgehog Review,* Spring 2019.

An earlier version of chapter 14, Kosher, appeared as "But Will the Lab-Grown Meat Be Kosher?" in the *Revealer: A Review of Religion and Media,* November 7, 2016.

Excepts from "Hollyhocks in the Fog" from *Before the Dawn/Hollyhocks in the Fog,* by August Kleinzahler. Copyright © 2017 by August Kleinzahler. Reprinted by permission of Farrar, Straus, and Giroux.

Excerpt from "Anonymous: Myself and Pangur" from *Poems 1968–1998,* by Paul Muldoon. Copyright © 2001 by Paul Muldoon.

_____

ONE

Cyberspace/Meatspace

I wake up to a weird future. It is 4:30 A.M. in Los Angeles on August 5, 2013. I'm about to watch the food of tomorrow appear at just past noon in London, my bleary eyes and smudged computer screen a double set of windows into space and time. I set my browser to www.culturedbeef.net. The future will arrive in the form of laboratory-grown meat made of bovine muscle cells that proliferated in a bioreactor. Or at least that's how the press event I'm awake to watch has been billed. Each announcement has been filled with promise: meat will never be the same, nor will we.[1] A basic fact about humans is that one of our food sources has, for longer than we've been *Homo sapiens,* come from the bodies of dead animals. That might soon change, as technological progress moves us further along a track that leads from hunting to farming to the laboratory. Such transitions are serious business, but if we're perched on one of history's great pivot points, it's good to keep our sense of humor— there is something inherently silly about the idea of an international media event staged around a hamburger, one of the world's most recognizable, mundane, and American foods. At the world's fairs and expositions of a previous era, grand events that one critic called "sites of pilgrimage to the commodity fetish," novel foods were displayed to crowds of visitors inside glass pavilions.[2] I'm getting ready to watch the early twenty-first-century equivalent, coffee mug clutched tight.

Journalists have described the hamburger in question as a "frankenburger," "test-tube burger," or piece of "vat meat." It was produced not by killing and butchering a cow, but through the expensive and laborious use of a well-established laboratory technique known as tissue or cell culture, first accomplished by the American embryologist Ross Harrison in 1907.[3] After decades of use in scientific and medical research, tissue culture has only recently been used to produce what is sometimes called, with technical accuracy but zero gastronomic zest, "in vitro meat." One of the many promises attached to this new meat is that it could offer an alternative to industrial animal agriculture, perhaps completely replacing its environmentally damaging and cruel practices with pacific ones. This meat's utter weirdness cannot be overstated. Meat that never had parents. Meat that never died (in the sense that a whole animal dies) and, in the eyes of some critics who define their meat narrowly, never properly lived. Meat that could utterly transform the way we think of animals, the way we relate to farmland, the way we use water, the way we think about population and our fragile ecosystem's carrying capacity of both human and nonhuman animal bodies. A new kind of flesh for a planet of omnivorous hominids who eat more meat with each passing generation. As my Los Angeles neighborhood stirs in the early morning, cyberspace becomes meatspace.

Clickbait stubs have swarmed through the Internet in recent weeks, drawing bits of human attention (perhaps the Internet's real currency—I'm spending some now) by announcing the burger's shocking price tag: over $300,000. Rumor has it that a single wealthy benefactor in the United States has funded the Dutch laboratory that grew the cells and shaped them into muscle and then meat. Mark Post, the medical doctor and professor of physiology who created the burger, is the man of the hour, but media professionals coordinated this event, paid by Post's benefactor. Cultured meat is a technology still in development, despite the very established nature of tissue culture techniques; this is one of the reasons it costs so much to produce a small piece of meat. In the local language we might use to describe this technology, it is "emerging"—a metaphor regularly used to mark the phase when novel types of computers, energy generators, or medical technologies are devised or discovered, built or grown, eventually tested and licensed, promoted in the media and (with painful slowness, from the perspectives of their designers and investors) become available to consumers. "Cultured meat" is a term that is just starting to surface as of 2013, and Post's use of the term at this event may be an effort to replace the clinical-sounding "in vitro meat."[4]

The "emergence" metaphor casts the future as a kind of mist out of which concrete things materialize. I think of the signs by which we track emerging technologies: patents, investments, research grants, conferences, exploratory launches of products in specific markets, splashy front-page profiles of entrepreneurs in technology magazines. Before my own meat brain is properly awake it occurs to me that the emergence metaphor performs a curious sleight of hand by hiding human agency. It implies that a new technology comes toward us of its own accord, rather than being ushered into being by many hands, each pair with its own agenda. And for a given technology to emerge, there must be a public for it to emerge into. Someone must be watching, and they'll have their own ideas about the future. I've been trained by utopian science fiction to expect certain things from a future of spaceships, and dystopian science fiction has taught me what to expect from a future Earth devastated by climate change, but do I know what to expect from a future of vat-grown meat? I train my eyes on my monitor.

For a subjectively long stretch of time, "feed will start soon" is all my browser shows, but then the event begins with a promotional film. A gentle guitar chord strums in the background and the camera shows gulls diving down over waves. A house is perched over the ocean. We see a bucolic human coastal settlement, the architecture noticeably North American or European. We're in the immediately recognizable aesthetic mode of a nature documentary or a science program aimed at young viewers. The camera pans out over the ocean, showing a lighthouse. Over this a voice states, "Sometimes a new technology comes along, and it has the capability to transform the way we view our world." Post's secret backer is revealed. A quick cut to a headshot of the speaker shows Sergey Brin, cofounder of the major Internet search and product company Google, and thus someone with a unique perspective on the way technology changes worldviews. But why is a Silicon Valley billionaire, someone who made his fortune from a search engine that has become so ubiquitous that "to google" is practically standard English, getting interested in the future of food? A simple lexical shift will reveal one answer to this question; cultured meat may someday be food, but right now it is part of what investors in Silicon Valley, Brin's domain, often call "the food space," an area of enterprise and investment that links food production and supply, environmental sustainability, human health, and the welfare of nonhuman animals. The food space is one in which venture capitalists have been very visibly active in recent years. But the word "space" has narrower and more specific historical connotations, conveying not mere dimensionality but also an intimation of the frontier. Frontiers are places different human populations have gone, over the centuries, in order to extract resources.[5] Some have argued that without frontiers capitalism itself could not function, for capitalists need fresh natural resources and new opportunities for the profitable investment of capital.[6] From the standpoint of shareholders, Google doesn't produce value by providing free search functionality to billions of people around the world. It produces value by establishing a new frontier: extracting the resource of our search data (and many other kinds of data too), which it then puts to undisclosed but immensely profitable use—and it also sells advertising space, a chance to catch human attention that was originally directed elsewhere.[7] Meat is already in our money in many parts of the world, through a trace quantity of tallow in the lining that coats our banknotes. You might say that commodity meat and money are already "spaces" for one another, reciprocally linked through use and investment.[8] This is how cows become capital—they are counted head (*caput* in Latin; thus "capital") by head.

Brin continues speaking, and the scene dissolves from the birds and the waves to a close-up of his youthful face with a fringe of salt-and-pepper stubble, framed by the device known as Google Glass. This is a headset designed in California and built by the Chinese company Foxconn, with a tiny computer screen the wearer can look into, gazing at the Internet while they appear to be gazing at those around them. Itself an emerging technology, Glass was released to the public in February 2013, but it is rare to see anyone walking around wearing the very expensive Glass (the name reminds me of glass pavilions from world's fairs) except in such tech-centric places as Palo Alto, California, or the blocks surrounding the Massachusetts Institute of Technology. Brin's decision to wear Glass in the film underscores his role as a very wealthy ambassador from the future. Brin speaks of his efforts to find technologies "on the cusp of viability," capable of being "really transformative for the world" (more promises, I note, and his phrasing reminds me that cultured meat may soon be an investment opportunity), and then the scene changes again.

A new talking head appears. It belongs to the senior biological anthropologist Richard Wrangham, who sits in his Harvard office, book spines visible on shelves behind him. He's apparently here to explain the transformative potential of which Brin spoke. "The story of human evolution," Wrangham says, ". . . is intimately tied to meat." He proceeds to tell a common and widely shared story about the importance of meat in our species' natural history, a version of which is included in his 2009 book *Catching Fire: How Cooking Made Us Human.*[9] There Wrangham argued that our evolution into modern humanity was made possible by cooking, and especially by cooking tubers and meat, abundant sources of calories that facilitated the development of several features of our contemporary morphology and sociability: small mouths, large brains (the brain is a calorie hog), a facility for cooperation, and a distinctive social structure based on reproductive relationships between males and females. Wrangham's is a radicalized version of other tales about humans and their evolutionary relationship with meat and other foods. His book has been subject to discussion and debate among biologists and anthropologists in a way that the film I'm watching can't possibly track.[10] The tactical reasons for bringing Wrangham into the picture are clear. If Brin speaks for the promise of new technology, Wrangham speaks for evolutionary antiquity and the authority of science.

Whether one agrees with Wrangham or not, it's impossible to miss the way the film matches a story about our hominid past with a story about the future of meat. Why suture together the deep time of species identity and the shallow time of our imminent dietary choices and food-provisioning strategies? Is evolutionary antiquity supposed to ground and legitimate hypermodernity? Am I to think that the past justifies the future? The next sequence jars me out of such reflections, as we cut from Wrangham to a piece of meat being cooked over a campfire in the darkness. The meat is on a stick held by a long-haired human, naked save for a loincloth, features obscured by darkness and the glare of the fire. Then a quick cut to African tribesmen, carrying spears and running barefoot. Wrangham goes on: "Hunters and gatherers all over the world are very sad if, for a few days at a time, the hunters come back empty-handed. The camp becomes quiet. The dancing stops." Wrangham's voice grows more animated and he raises his fists: "And then someone catches some meat! They bring the prey into the camp"—the camera jump-cuts to a new, distinctly modern scene, in which an adult white male opens the lid of an outdoor grill—"or nowadays, into someone's back garden barbecue." The two registers, the stereotypical African-primitive and the white and modern, are suddenly fused to a specific purpose, as if to explain and justify Western and modern behaviors by reference to "primitive" ones. The move is familiar, and offensive though probably innocently intended. It's the sort of fusion that took place in the after-school science programs I watched as a child, or in some older nature documentaries; it comes as a considerable surprise to see such recourse to the notion of the primitive many decades later. It is the visual equivalent of what anthropologists have criticized as an unthinking sociobiological turn.[11] As the film continues, Caucasian children stare at modern meat in the form of hamburgers. Wrangham says, "Everyone gets excited to come and share. . . . It is ritually cut." A knife-wielding white male in a baseball cap divvies up a steak. "We are a species designed to love meat."

The symbolic assignment of modernity to Western white males, and of an ancestral past to black Africans, is surprising in a promotional film released to an international media audience in 2013. Yet Wrangham's claims hold a different kind of surprise. In less than a minute of exposition, Wrangham (as presented by the film's director and editor) has achieved a magnificent elision of meaning, moving from the idea that cooked animal flesh played a crucial role in producing human physiological and social modernity to suggesting that our taste for meat is original, innate, that it is natural for us to desire it. According to this logic, vegetarianism represents a break from our "design." But this logic is a tangle. The idea of a natural taste for meat is not uncontested, and this contest may in turn be the iceberg-tip of a deeper scientific debate regarding the status of humans in the food chain and our relationship with other forms of animal life. Technology is implicated in the practice of hunting animals, and thus our relationship with meat is linked to our status as tool-making and tool-using creatures. This latter point is not lost on cultured meat's advocates. Some of them argue that laboratory-grown meat may be a logical extension of our gradually changing and inherently technological relationship first with subsistence itself and then with industrial food production. "Designed to love meat" is a slogan that invokes hominin evolution as a license to pursue the love of meat in whatever modern way technology enables.

The film won't wait for me to summon footnotes to mind, of course.[12] It moves on to a conveyor belt carrying pink hamburger patties directly into the camera lens. We've dropped the question of human appetites and picked up the crucial question of scale, announced with this look into the guts of our industrial meat production system. A new expert, the environmentalist Ken Cook, says, "Feeding the world is a complex problem. I think people don't yet realize what impact meat consumption has on the planet." With a quick cut to cows in a field, Cook and Brin alternate to provide a few bullet-point problems associated with industrial-scale animal agriculture, the problems that cultured meat's pioneers hope to remedy. For example, 70 percent of antibiotics used in the United States go into livestock bodies, not human ones, and those antibiotics are required partly because of the cramped conditions in which livestock are raised and kept before the slaughter.[13] Another important reason for the use of subtherapeutic doses of antibiotics is that it enhances the rate at which animals put on weight, bringing them to slaughter faster. "When you see how these cows are treated . . . that's certainly not something I am comfortable with," says Brin, reminding me of the obvious problem of animal ethics, but the other side of such intensive antibiotic use is that it has been known to breed antibiotic resistance in the pathogens that circulate among livestock. This makes concentrated animal feeding operations (CAFOs) breeding grounds for viral agents dangerous both to livestock and to humans. Stories about the hazards of CAFOs and slaughterhouses have become commonplace. From a dystopian perspective the "future of meat" isn't lab-grown meat, it's a global pandemic originating in abused and crowded animal bodies.[14] Cook reminds us of the health risks associated with simply eating a lot of meat; high levels of carnivory are associated with a 20 percent greater-than-normal chance of developing illnesses such as heart disease or cancer. However, as I will come to learn, more supporters of cultured meat are motivated by the next issues he raises: the environmental cost of meat production, which is thought to yield about 14–18 percent of industrial society's greenhouse gas emissions annually, and which uses an enormous amount of water and land. These resources could feed more mouths if they were devoted to fruits, grains, and vegetables instead. In 2011 a graduate student at Oxford conducted a theoretical life-cycle assessment of cultured hamburger, comparing it to the conventional kind. While the assessment favorably compared the lower environmental costs of cultured meat production to those of conventional meat, it was also declared full of holes by critics and was eventually revised.[15]

More images of farmland, then a runner passes in front of the camera while Cook describes the healthier diet of the potential future. Then we cut quickly to the crowded streets of Amsterdam's central neighborhoods, with their canals and bridges, and Cook gets to the heart of the issue, our growing global population. He expresses an idea that I will hear often as I make my way through the cultured meat movement, namely that meat consumption is rising faster than population growth alone explains. Some expect global carnivory to increase by 50 percent by 2050. I blink, noticing a prediction being taken for granted as if it tracked a natural law. "We're in for a terrible reckoning," Cook says as the camera cuts to a field, dust rising in the wind. This is grim, but predictions for an increasingly carnivorous humanity have substantial precedent. Meat consumption doubled worldwide between 1960 and the 2010s, and it increased even more, and faster, in later-modernizing countries like China. Wrangham's voice returns, reminding us of the pressing problem of climate change, which promises to collide with population growth, shifting resource distribution in ways that will promote conflict. "In a modern world, where we have Paleolithic minds [I choke on this a little] and contemporary weapons, that's really dangerous." Wrangham has returned to his strange fusion of the modern and the prehistoric, invoking Paleolithic minds (he probably means brains that became effectively modern in the Paleolithic—that is, prior to the technological and agricultural revolutions of the Neolithic) as if cultural change and modern civilization matter little when it comes to the basics of human behavior, as if the mind is not much more than the meat brain with its meat instincts. But the film has also smuggled a prediction inside a prediction: if we don't develop technologies to thwart resource scarcity before widespread crisis hits, we'll be savages playing with nukes.[16]

Implicit in Wrangham's image is yet another intriguing but questionable idea, one that I will encounter repeatedly in the course of my travels within the cultured meat movement. This is the idea that the modern human condition is constituted by a disharmony between our biology and our technology, a lack of synchronicity between our bodies and their myriad artificial extensions. Everything about modern meat returns us to this notion of disharmony. We maintain a polluting, dangerous meat production system, a form of artificial infrastructure that allows an unprecedentedly large human population to consume an unprecedentedly large amount of animal flesh per capita per annum. This is not identical to the idea of a "machine in the garden," a technological presence that disrupts both the natural world and a sense of human connection to nature.[17] It is, instead, the desire to rediscover our biological condition from within the "second nature" that we have built around our bodies, and with which our bodies constantly interact, and to ask how that condition might be better served. The idea seems to be that our problems would be solved if we had better prostheses.

Once the viewer has been thoroughly exposed to the links between meat, population growth, climate change, and our dangerous future, Brin reappears to suggest that we might "do something new." A grassy hillside dissolves, and in its place we see a lattice of white lines over red, like a bird's-eye view of an organic city planned as a grid. This is in fact a close-up of animal muscle. Post's voice rises above it: "By our technology we are actually producing meat. It's just not in a cow." Post identifies himself as a physician experienced in vascular biology, with the goal of creating tissues for human transplants, especially blood vessels for heart patients. Referring to the fact that stem cells—unspecialized cells that can replenish themselves via cell division and, either in bodies or in experimental media, become cell types that fulfill specific functions—have been seen as promising for the production of human parts intended for transplantation, he says that "stem cell techniques are very useful for growing beef." My monitor has grown dark, but a cluster of red cells glows in its center, a model that will illustrate Post's process. "We take a few cells from a cow, muscle-specific stem cells that can only become muscle." A single cell divides, an animated exemplar that resembles a celestial body floating in the void. Post continues: "There's very little that we need to do to make these cells do the right thing." He describes the way muscle cells proliferate and divide, creating functional structures almost all by themselves. Via technology, we simply provide anchor points and future muscle fibers will form. "A few cells that we take from this cow can turn into ten tons of meat."

Post's remark reminds me of the 1952 science fiction novel *The Space Merchants,* by Fredrik Pohl and Cyril M. Kornbluth, in which an entire factory of workers is fed by a giant, quivering, gray hemisphere of chicken flesh called "Chicken Little," whose creaturely status is uncertain; she lives on algae and occupies a nest in the basement.[18] Post's statement also recalls the scientific and medical discourse that has emerged, in the last two decades, around the stem cell, that enigmatic but ever-present object of hope, about which news items appear each week.[19] Post's "ten tons of meat" is just one of the miracles stem cells are expected to perform. Others range from regrowing broken teeth to reducing the physiological age of human tissue. In the world of cardiology, Post's world, stem cells are expected to yield therapies that add years to patients' lives, therapies that would also (needless to say) represent a source of immense wealth for the medical industry: here stem cells offer both economic and biographical potential.[20] Running through all of this are the complex dynamics of promising; like some other observers of biotechnological hopes, I am reminded of Friedrich Nietzsche's observation that humans are defined by their status as creatures who make promises. Nietzsche's specific claim was that, in this regard, humans are a "paradoxical task Nature has put to itself."[21]

Gentle music strikes up. The sun rises red in a red sky above red hills. This could be a science fiction film, or California reddened by airborne particulate matter (I will later learn that the Department of Expansion, the documentary film company that produced this film, is based in Los Angeles). We hear Brin's voice again: "Some people think that this is science fiction, that it's not real, it's somewhere out there. . . . I actually think that's a good thing. If what you're doing is not seen by some people as science fiction, it's probably not transformative enough." A quick cut: a man's hands (white ones; I realize I've become race-conscious because of the earlier juxtapositions of Africans and Europeans) drop some hamburger meat from wax paper onto a wooden surface, where they mold it into a patty. Brin: "We're trying to create the first cultured meat hamburger. From there I'm optimistic that we can really scale." To pause on that crucial word "scale," here used as a verb, the price tag of the first cultured beef burger reflects ample research-and-design time, the salaries of technicians, plus expensive laboratory supplies, and it benefited from no economies of scale—it is massively higher than the potential (that word again) cost of the burger at scale. Such talk of potential brings us back around to the ultimate target of the cultured beef project, the future. Post again:

> Twenty years from now, if you entered the supermarket, you would have a choice between two products that are . . . identical. One is made in an animal. It now has this label on it [stating] that animals have suffered or have been killed for this product. And it has an "ecotax" because it's bad for the environment. And it's exactly the same as an alternative product that is being made in a lab, it tastes the same, it has the same quality, it is the same price or even cheaper, so what are you going to choose?[22]

As he speaks, we see images of children and their parents happily munching on hamburgers. "From an ethical point of view, it has only benefits."

As Post continues, our scene shifts from the burger-munchers to an arboreal display that could only be Northern Californian. We look up at soaring redwood trees from their bases, viewing an environmental treasure whose preservation is one of the "ethical benefits" of which Post speaks. Water drips and minnows swim, as Cook describes growing consumer interest in new systems of food production that may not damage the environment. Then we return to Wrangham, who speaks of meat's benefits as he did before, but with a difference: "Now, by some horrendous irony, it's become part of a system that threatens our species. We have to do something about it." The image of Wrangham in his office fades to a white screen on which the words "Be Part of the Solution" appear in black letters.

Environmental crisis. The unstoppable power of human appetites. Flesh, both the flesh we eat and the press of human bodies in our crowded cities. And against the onward rush of the disasters of climate change and population growth, another trend line, a more hopeful one accelerating upward, labeled "technological progress." The six-minute film is almost too much to take in, a kind of signifying fire hose, but it lays out many of the puzzles that will preoccupy me for the next few years as I quest after the meanings of laboratory-grown meat. This isn't a mere product demo that I'm squinting at over the Internet, it's an effort to position cultured meat as a new food technology that can resolve a problem whose scale is civilizational, so large that any effort to calculate it requires the tools of social and environmental science. Spaceship Earth's problems can be seen from orbit.

And while the film did not say so explicitly, it seems clear that the core problem's name is not exactly meat itself, however much conventional meat production is an important immediate target of criticism. Just where the problem lies, however, is ambiguous, and the film raises questions about our civilization that are too large to easily grasp but that demand more than mere hand waving. While much of the film locates the problem in that strange quantum called modernity, Wrangham's contributions are more troubling, inviting us to view human appetites as fundamentally at odds with our species' survival. Meat makes and unmakes us, according to the narrative toward which Wrangham gestures. Or is it sheer civilizational scale that makes and unmakes us? Is it technology? And what would it mean for modernity if technology can save the same natural world that it imperils, or more cynically, what does it mean that some people have become convinced that one technology can undo the problems created by another? And how would those sentences read differently if the word "technology" were replaced by the word "capitalism"? What if the solution lies not in producing more, but in needing less, and in the more just distribution of what we already produce?

And, if the future is coming in the form of tissue-cultured animal muscle to be consumed as meat, what does it mean to wait for it? The promotional film is true to the style of thinking that accompanies cultured meat in its early, "emerging" years. This style is hopeful, worried, sincere, and immensely ambitious, responding to a grandiosely scaled map of the world's problems that its proponents have themselves drawn up, a map that usually leaves out the basically political character of those problems, just as the metaphor of emergence slides past the tangle of political and financial interests out of which new technologies actually emerge.

Now my screen shows the interior of a television studio, full of journalists. There is a modern kitchen counter and a small stovetop. A host welcomes Post to the stage, which is set up as if for an anonymous cooking show. They chat briefly, and then it's time to unveil the burger itself. Post lifts the lid off of the tray and reveals the burger, which looks very pink; it's been colored with beet juice and saffron, without which it might be a muted white-gray. Insofar as visual inspection can reveal texture, it appears to be very different from conventional meat, and we are told that it has been thickened with bread crumbs. A chef named Richard McGeown and two other guests then join Post onstage. One is Josh Schonwald, an American food writer with a book on "the future of food" to his credit, and the other the Austrian nutrition scientist Hanni Rützler.[23] The chef receives the burger at the stovetop and uses just a little vegetable oil and butter to cook it, as the camera moves between close-ups of the stovetop (it must be a little nerve-racking to cook such an expensive piece of meat) and the expectant faces of members of the audience. The burger does indeed start to brown in the same way that conventional meat does when the Maillard reaction begins.[24]

Later I'll learn more about why Post, an amiable, tall Dutchman who speaks the excellent English of an educated European who has lived in the United States and travels often, chose London: every major media outfit has a London bureau or roving journalist, and Greenwich Mean Time still enjoys a certain global centrality. I'll also learn that Post's team would have had more trouble getting its hamburger past the U.S. border than past the British one, a surprising detail because the British are understandably—given prior outbreaks of bovine spongiform encephalopathy ("mad cow disease") among British cows—touchy about meat. The lab-grown hamburger isn't just a new form of meat; it is also a border-crossing alien, albeit a legal one. I wonder what all this means for the eventual regulation of cultured meat as a food product.

While the burger cooks, Post shows us a second film, an animation illustrating the process by which he and his team produced their burger. A tiny biopsy of muscle tissue was taken from a cow, which was barely grazed by the experience, and returned to grazing. After skeletal muscle stem cells were isolated they were encouraged to proliferate in a growth medium. As the cells grew they were encouraged to form chains, strands that would later be turned into the muscle tissue of hamburger meat; those strands were "exercised"—in other words, encouraged to expand and contract as skeletal muscle does in vivo. I know enough about tissue culture to suspect that the process was somewhat more complex than this. It certainly was time consuming, since it took several months for Post's lab to grow enough material to produce their burger.

McGeown finishes cooking the burger, which he describes as having a "very pleasant aroma." He turns it out onto a plate along with a tomato slice, lettuce, and a bun, although he doesn't assemble the burger to be held and munched. It sits in the center of the plate, naked, as if contesting the historical role the bun played in defining the qualities of a hamburger sandwich. Each of the two "taster experts," Schonwald and Rützler, cuts into the meat with a knife and fork and samples some that way. Both report that it definitely does not taste like conventional meat, but Schonwald attests that it reminds him of the "mouthfeel" or "bite" of meat. Post takes a bite himself.

Cultured meat apparently eats like real meat, even if it doesn't taste exactly like it. Throughout this entire process, the studio audience of journalists has been visible, and now they're stirring, impatient to ask questions. Post is ready to field them, and the first two are critical. The first: Will consumers *want* to eat meat made under laboratory-like conditions? Post acknowledges that there's a powerful initial "yuck factor" that we need to bear in mind, a potential resistance to meat that wasn't grown in animals. I'll encounter this during my research in the form of a hard psychic line drawn between the kitchen and the laboratory, as if much of our food hasn't already passed through institutional kitchen-laboratories shaped by science.[25] The second question from the audience is about whether a new source of a large volume of meat would encourage people to eat more meat than a healthy diet suggests. Post nods, understandingly, and says that he himself is a "flexitarian" and would happily see us all eating less meat. However, he goes on, the hard truth is that meat consumption will only continue worldwide; the "meat question" will not be resolved by mass vegetarianism or flexitarianism. And Post continues to respond, in the same open spirit, to a long series of questions, many of which target apparent weaknesses or flaws in his plans. Post acknowledges that his techniques are at an early stage of development, currently too inefficient, scarcely near the point of "scaling up." Furthermore, a replacement for the current growth medium must be found. That medium includes fetal bovine serum, making the whole process emphatically nonvegetarian, and moreover, antibiotics have been added to the cell culture to prevent a damaging infection. One solution to the problem of overreliance on antibiotics, Post says, would be to use robotic and thus totally sterile production facilities.

To an additional question about the burger's taste, Post responds that his team has not yet mimicked the taste and mouthfeel of animal-grown muscle tissue. One reason for this is that they have not yet learned to generate the fat cells such tissue would contain. Not only does fat contribute to flavor in many ways, it also adds much to our sense of meat's tenderness.[26] The popularity of lean cuts of meat among health-conscious eaters should not obscure the central role played by fat, even small amounts thereof, in creating the taste of meat. As he addresses question after question, Post remains optimistic and upbeat. Asked whether cultured meat would start rolling off assembly lines in a week, and onto the shelves of Sainsbury's (a British supermarket chain), he laughs appreciatively, as he does at questions like "How much will it cost?" Today's demonstration was strictly a proof of concept, and Post limits himself to the conservative prediction that cultured meat may not be available for another ten to twenty years. I pay careful attention to this. Such predictions have appeared in the media with striking frequency as part of the media swarm around this event, and one journalist even takes the time to assemble them, creating a chart entitled "When Will We Eat Hamburgers Grown in Test-Tubes?"[27] I am not the only watcher conscious of the way cultured meat is bound up with a culture of prediction, and of the relationship between the long timeline for perfecting Post's technology and the possible funding streams that might support it. Isha Datar, who heads an organization called New Harvest, founded in 2004 to promote research into laboratory-grown meat (Post's lab is not the origin of this technology; he is just the latest and best-funded entrant into the field), brings up an interesting point about how such work gets support. At the moment, the money for vat meat research is principally philanthropic, because the venture capitalists who support companies need to see returns in far shorter increments of time than twenty or ten years. I expect this to change as cultured meat develops an aura of viability through demonstrations like this one.

A Brazilian journalist, in a jocular tone, voices his doubts about whether you could produce a good barbecue using laboratory-grown meat. Post agrees that to truly replicate meat is a huge challenge. Tastes are complex. There are some four hundred peptides and aromatics in meat, and no food scientist can tell us exactly how the composition of meat yields specific tastes. For a moment I think that the question-and-answer session will end on this relatively gentle and optimistic note—a scientist working to complete a very difficult but not impossible task, with the fruits of this labor helping to resolve civilization-scale challenges. Instead the last word comes from an audience member who expresses her irritation at Post's failure to bring enough to share with the whole audience. This too is greeted with laughter from the room, and the event ends. Watching through my Internet browser, I find that I cannot blame her for wanting a bite. After all, in the twenty-first century we are bombarded with images and words designed to summon the future. Rare is the chance to engage with the future through the intimate senses of taste, smell, and touch.

In the years of this book's research, from 2013 to 2018, I went out to find the lineaments of my larger society in the concepts of its speculative biotechnology.[28] Cultured meat was not just an emerging food technology. It was an emerging conversation, a climate of opinion condensed into a physical object—in fact, into a very small physical object, because between 2013 and 2018, no cultured meat was being produced beyond the level of small tests such as Post's burger. The charismatic pull of that conversation has been enormous, though, and for good reason. It is a conversation about what our world might become. It has linked human actors ranging from vegan activists in Brooklyn to designers in Amsterdam, venture capitalists in San Francisco, and biohackers in Tokyo, not to mention laboratory scientists from a wide range of disciplines and a handful of social scientists, journalists, writers, and professional futurists (or "futures workers," as these consultants are often known). Everyone brings their own desires to the subject; there are, of course, entrepreneurs who desire wealth and fame as well as those for whom entrepreneurship is a means to an end. There are activists who hope to set food animals free, and others who want food security for a growing population, or to mitigate climate change, and there are scientists pouncing on a technical challenge. Meat's meanings are multiple, and this holds true for the lab-grown kind, too. There are also gadflys who believe, despite Post's burger, that cultured meat can never work, that Post and his colleagues will never find a way to scale up to industrial production, leaving cultured meat nothing more than a novelty of its time, the biotechnology equivalent of a giant elk whose antlers are outsized for its survival.[29]

This book tells the story of what I found, and what I did not find, in the course of my time in the small, strange world of cultured meat, during what seemed to be the early years of an emerging technology. I expected to spend time in laboratories, observing scientists and learning how they encouraged cells to proliferate, and exploring their expectations for the future of cultured meat. This did happen in some measure, but for the most part I found myself with very little laboratory science to observe and a great many public conversations about cultured meat to participate in and sort through. During my five years of research, the world of cultured meat changed dramatically, fed by venture capital, media interest (an inevitable pun: cells feed on growth media while an embryonic industry sometimes thrives on attention from the media), and the growth of more than one nonprofit organization devoted to promoting cultured meat and other technological alternatives to animal agriculture. At the very beginning of my research, there was only Post's burger and an expansive, and perhaps unanswerable, set of questions about what would happen next. In other words, we were in the territory of professional futurists who speculate about where new technologies might lead us, and accordingly I spent time with futurists in the consulting firms and non-profit organizations that are their workshops. Anthropological fieldworkers have traditionally learned local languages after reaching their field sites, out of necessity. I busied myself by reading the small scientific literature on cultured meat, and by speaking with entrepreneurs and investors to learn the idioms in which both science and investment articulate their goals. As of 2013, the most commonly asked question was "When?" or "How soon?" and the answer given by most researchers and observers of the field was "About ten years"—ten years until a marketable cultured meat product could reach consumers, perhaps beginning the process by which cultured meat would undermine conventional animal agriculture.

I quickly learned that Post's hamburger had emerged out of a small world of cultured meat researchers who preceded him. Around the turn of the millennium, a grant from NASA had funded a team at Truro College in New York, led by Morris Benjaminson, that attempted to turn goldfish cells into a compact and self-replenishing food source for long space flights. Meanwhile the artists Oron Catts and Ionat Zurr were using fetal sheep cells to create "living sculptures" from a lab at Harvard Medical School. Post himself was originally part of a consortium of Dutch researchers operating with a substantial grant from the government, won through the persistence of a Dutch businessman named Willem van Eelen. In other words, the potential for tissue culture to produce cells for nonmedical applications was apparent to a range of actors with different purposes. During the first decade of the twenty-first century, all this work unfolded in relative quiet. People for the Ethical Treatment of Animals (PETA), seeking to catalyze research, announced a contest in 2008: the first laboratory that could produce a chicken nugget made via cell culture would win a million U.S. dollars. No one collected, but PETA did make the papers.

What quickly coalesced in 2014 and 2015, perhaps catalyzed by Post's burger demonstration, was a climate of eager conversation about cultured meat and the future of food in which elites from developed nations, most especially the United States, the Netherlands, and Britain, discussed the possibility of feeding the world through a new subsistence strategy. This strategy would be in keeping with these elites' ideological preferences, organized (as in Post's demonstration) around environmental protection, sustainable protein production, animal welfare, and human health. A group of actors from biomedical research, venture capital, the nonprofit world, and other fields unselfconsciously played a role that other elites have played over the past two centuries of European and North American history. They cast themselves as food planners for the globe, and arbiters of proper dietary practice for both the well fed and the poorly nourished.[30] This role playing, which arguably goes back to Thomas Robert Malthus's *Essay on the Principle of*

Population (1798) and whose original political context was British colonial expansion, retains its political character, even when this is not explicitly acknowledged. The preference for solving problems using technology is very often a political preference even when it appears to ignore politics.

The developments that actors like Post ponder and debate are very real, and they include agricultural land becoming unusable (or even flooded) due to climate change, the effects of rising global temperatures on the bodies of farm animals, and the possibility that rising middle classes will consume more and more meat. But their proposed responses reflect specific beliefs (Western ones) about what constitutes a desirable human diet, and beliefs about the right relationship between human eaters and the ecosystems from which their food comes (industrial ones). I was in the world of cultured meat as a kind of anthropological fieldworker, but I was also led into the deeper histories of the debates I witnessed, and this book is as much a work of history as it is an ethnography (to use that strange term, which literally means "the writing down of a people"). "Soylent Green is people" runs the tagline from a classic dystopian film about the future of food, in which green wafers are made to feed a population that has grown beyond the limits of sustainability much as Malthus once warned it would, and those wafers are made from reclaimed corpses. Cultured meat, conversely, is not people, but it rests on a series of claims about the human condition, both in its physical aspect and in the sense of what we consider a good life. That banal phrase, "a good life," becomes more meaningful when shifted into the idiom of philosophy. What is a good life, one in keeping with our ethical beliefs about purpose, dignity, and posterity?

When my research ended in 2018, much had changed. Post was still one of the leading figures in the field, but in a new role as one of the founders of a company, Mosa Meats, to which the 2013 hamburger demonstration would subsequently be attributed. Hampton Creek, a company previously known for vegan mayonnaise, suddenly revealed that they had been working on cultured meat and promised to put some on customers' plates (it was not clear which customers, where, or what the meat would be like) by the end of 2018 (by which time Hampton Creek would take the new name "Just"). Memphis Meats, which despite its name is based in the Bay Area, had unveiled samples of chicken strips and pork meatballs, two types of meat that, like hamburger or sausage, are less dependent on texture than steak would be. There are other players in the "space" too, making their own promises. What 2018 and 2013 shared was a focus on the question of "when," but the existence of specific players making ambitious promissory statements changed the dynamic considerably, as did the inevitable black boxing of research. It was possible for a visiting scholar to get inside a start-up's in vitro meat lab in 2013 or 2014 or even 2015, but this had become very difficult by 2018. This meant that even as the companies seemed to make progress, the ability of social scientists and journalists to confirm that progress diminished. My research began in one kind of fog bank and concluded in another. Tracking emerging forms of biotechnology can make you cynical, but part of what is at stake here is our capacity for sincerity in the face of grand challenges. Sincerity is complicated when one does not know whom to trust or believe.

At one point during my fieldwork within the cultured meat movement, "the post-animal bioeconomy" became a buzz-phrase of sorts, used to describe a range of techniques, often involving tissue culture, for developing products humans have traditionally obtained from nonhuman animals. Such a phrase bespeaks a lot of ambition to say the least. It would take an effort far beyond that of a few coordinated start-up companies, consultants, and promoters to make our "bioeconomy" truly "post-animal." The post-animal bioeconomy, even if it is still a matter of the imagination, is intertwined with another kind, a "promissory moral economy." In these intertwined economies we invest hope, energy, and attention in novel technologies that are moral in a double sense: not only would these technologies have desirable moral outcomes (particularly from animal protection perspectives), but they function as ways to express moral feelings even before the desired technology emerges. To support cultured meat is, for many, to condemn CAFOs and perhaps all animal agriculture. Such expressions bring activists together and justify the use of the term "movement" for the effort to bring cultured meat to pass. We watchers, especially those of us with our feet planted in history or anthropology, are often suspicious of promises coming from the world of technology. Indeed, an "ethic of suspicion," as the historian and anthropologist of genetics Mike Fortun has called it, has become central to the way we watch.[31] It is a curious thing to meet a moral economy with an ethic of suspicion, but such encounters are common as world-saving claims are made on behalf of emerging technologies that arrive with business interests attached.

Cultured meat was a glittering object in the media during my years of research, but a holographic one, without solidity. News articles vastly outnumbered researchers and laboratories. To the best of my knowledge, then and now, very little cultured meat has been produced, and nothing beyond the scale of Post's 2013 hamburger. But the relative absence of much cultured meat in those years is precisely the point. Cultured meat was, and remains as of this writing, a technology that has not fully emerged, and thus remains largely an abstraction. If this book reads like a series of detours—"Where's the meat?" the reader may ask, and it's a reasonable question—it is because my research often took the form of detours and delays. This was frustrating at first but later became interesting, because what I found "on the way" to cultured meat was a set of questions whose intrinsic intellectual worth is great. Unexpected detours, one might say, are the opposite of predicted routes, and thus the opposite of a certain style of futurism too, the kind that asserts the knowability of a particular future, often a future presumed to follow from the development of a particular kind of technology. Detours turn a planned journey into a series of surprises, perhaps pleasant, perhaps regrettable. For me, the detour starts out as an irritation or a disappointment. Then it becomes a method. This book's arrangement of chapters is the result of that method. They move between past and contemporary frames of reference, between concerns that are anthropological and historical and philosophical. They contain very few hard-and-fast answers to concrete questions such as "Will cultured meat succeed?" and "When is cultured meat likely to arrive?" and "What does it taste like?" This is not only because those questions are, as of this writing, without final answers, but also because I contend that they ultimately matter less than the questions this book does ask, the essential one being "What makes cultured meat imaginable?"

This book is not an attempt at prediction but rather a study of cultured meat as a special case of speculation on the future of food, and as a lens through which to view the predictions we make about how technology changes the world. Almost all of those predictions, whether made by professionals in consulting firms or think tanks, by scientists and entrepreneurs with a personal investment in the work, or by members of the general public, have been influenced, at least to some degree, by science fiction, that ubiquitous form of lay futurism. As of this writing, cultured meat is still an unwieldy bricolage of communications, a holograph projected from no point in particular.[32] It is often described as a sign of the gradual triumph of science and progress over civilization's ills, but it is more like an engineering project at whose center passions and interests churn. These range from a heartfelt desire to eliminate animal suffering to sheer cupidity.

Still waking up to a weird future in 2013, I have no idea about this yet. The hamburger demonstration ends and I close my computer, crossing from cyberspace back to meatspace.

———————

TWO

Meat

Protein is protean. The word "protein" derives from "the first" in Greek *(protos),* and it is as changeable as Proteus, the sea-god Poseidon's eldest son. This is perhaps the most important thing to know as we turn to the question of meat's definition, and to meat's shifting role in the human diet, over periods of time so long that they stretch from recorded human history back into natural history. Cultured meat was announced with a boast of novelty in 2013, and justifiably so. In comparison to conventional meat, cultured meat would mean a revolution. It would require new kinds of production facilities, methods, and tools. It would move through an entirely new food production infrastructure. It would demand stainless steel, glass, and plastic in great quantities, likely assembled into giant bioreactors resembling industrial-scale beer-brewing tanks. A novel bioeconomy, with new investors and new financial winners and losers, would grow and exert its influence. There are precedents for the mass production of a consumable via cell culture, perhaps most importantly Jonas Salk's creation of a mass-producible vaccine for polio in 1952,[1] but a vaccine is hardly a straightforward precedent for cultured meat. In the former case, a minuscule byproduct of cellular metabolism is gathered and used. In the latter, the cells themselves, and more to the point, organized masses of billions of them, are gathered and shaped for consumption. There is an important distinction to strike between products that are made of cells and products that are made from the byproducts of those cells' life processes.

If cultured meat were adopted so enthusiastically that its architects' dreams were achieved, and it actually began to displace conventional animal agriculture, the earth's animal biomass would be transformed in the process. The bulk of that biomass consists of domesticated creatures living and dying in our food system. The geographer Vaclav Smil has estimated that, as of 1900, some

1.3 billion domesticated large animals existed on the earth. As of 2000, the live weight of domesticated animals had increased some 3.5 times. Memorably, Smil imagines "sapient extraterrestrial visitors" concluding, on the basis of the sheer abundance of one creature in particular, "that life on the third solar planet is dominated by cattle."[2] If cultured meat suddenly replaced its conventional antecedent, billions of gregarious vertebrates would become unnecessary, their fates uncertain, much like the fate of the land used to feed and house them, and the water consumed by their care and processing, not to mention an entire industry and its workers. The suffering caused by industrial animal agriculture would end, replaced not so much by sudden relief as by a question mark. Some 75 percent of the earth's agricultural land is currently used, either directly or indirectly, for animal agriculture that produces meat, dairy products, and eggs.[3] That land, too, would become a series of question marks. The critic John Berger once called the zoo an epitaph to a lost relationship between humans and animals.[4] Our feedlots and slaughterhouses are, in a very different way, also epitaphs to lost relationships, ones likely beyond recovery.

Yet for all its novelty, cultured meat emerges out of a set of older, preexisting ideas and practices around meat, a set of linked histories of carnivory that are difficult to access from the standpoint of Mark Post's 2013 hamburger. If Post's burger represented the sum of our knowledge about meat, it would be impossible to work backwards from that starting point and reconstruct the history of humans eating other animals. Such a thought experiment might run from hamburgers made out of cows, in an industrial, fast-food format, back to the original European hamburgers of the mid-eighteenth century (sometimes called "Hamburg steaks" by English cookbook authors),[5] which were produced in a world of preindustrial meat. Our experiment would quickly reach periods when humans ate species that have disappeared as food animals, in many parts of the world. Swans, for example, no longer grace the tables of European elites.[6] If we recognize that meat has changed many times, and for many reasons, it sheds a new light on the modern, Western habit of using the word "meat" to signify solid fact, or bedrock reality, or the most salient issue at hand. It makes that habit seem curious.

Perhaps the most striking thing about the English-language word *meat* is that, while its casual contemporary use often carries a sense of stable meaning, its history is one of changing usage. The first *Oxford English Dictionary* attribution for the word is from 900 C.E., when *meat* referred to solid food in Old English, as opposed to drink (we can observe a similar shift in the French *viande*). In its Old English spelling, *meat* was *mete,* derived from the proto-Germanic root word *mati* and linked to a variety of other words in the same language family, such as the Old Saxon *meti* and the Old Norse *matr,* or the Gothic *mats,* simply meaning "food." It was around 1300 C.E. that *meat* began to refer to the flesh of animals, in distinction from other solid foods. Before this, but after the Norman conquest of Britain in 1066, a split terminology took shape, drawing from both French and Old English, so familiar to contemporary English speakers that we rarely notice it. Whereas the Old English way of describing meat would be to say, for example, "meat of cow," the French would refer to *boeuf,* or *beef* (*mutton, veal,* and *pork* all derive from the French, too). In his 1825 novel *The Betrothed,* Sir Walter Scott offered an interpretive gloss on this difference, suggesting that the French-speaking Normans got their meat at one more level of remove from the slaughtered animal than the Britons, who often roasted their animals whole. The Britons couldn't help but keep meat's creatureliness in mind. Also notable, less for its relevance to the definition of *meat* than for the ancient links between carnivory and economic thinking, is another etymological link: the Old English *ceap,* meaning "cattle," gives us the modern *cheap. Ceap* also meant "property," a reminder of economies of direct exchange, and of ages when animals were commonly used units of value.[7] *Cattle* itself is also linked to *chattel* and once referred to any property, not just the four-legged kind.[8] Thus, early twenty-first-century earth is dominated, in terms of biomass, by living property.

Our modern usage of the word *meat* betrays little of the indeterminacy and flexibility previously carried by a term that once meant "item of solid food," though in some turns of phrase, such as "nutmeat" or "sweetmeat," we can hear echoes of that old meaning. What once indicated solidity and edibility now means the muscle and fat of killed and butchered animals, exclusive of their *offal,* a term that derives, Germanically, from *ab-fall:* that which falls away in the butchering process. In modern Rome, this is the "fifth quarter" of the animal, the *quinto quarto,* which takes its name from a premodern system of meat division in which the first four quarters of an animal, in order of descending quality, would go to the nobility, the clergy, the middle classes, and the military, respectively, leaving the "fifth quarter" for the peasants. The semantic shift of the word *meat* tracks closely, across modern European and North American history, with our ever-narrowing sense of what counts as meat, but old potential meanings do not simply die when they fall out of use. Plans for cell-cultured foods could suggest a return to earlier meanings of *meat:* solid food of any kind, not necessarily cut from a carcass. Or at least, there is a strong desire for that re-broadening among those scientists, entrepreneurs, and activists who celebrate "alternative" proteins—not limited to cultured meat and often including plant-based meat substitutes, as well as entomophagy, or eating insects.

It is striking that the first famous piece of cultured meat was a hamburger. Post's lab considered making a sausage, a form of meat more noticeably Dutch and, in other European contexts, linked with artisanal production[9] (though sausage meat and hamburger are cousins), but the international appeal of the hamburger won the day. The hamburger is an appropriate avatar for modern meat, associated not only with abundance, but also with industrial production, uniformity, speed, flexibility, and, frequently, with automobiles and drive-through service. Beef has powerful associations with Great Britain, but hamburgers are American, and the hamburger is a particularly American symbol of plenty.[10] The bun, specially designed for the patty, is what made the sandwich edible by hand, on the go: fast meat.[11] One of the many ironies of the story of cultured meat is that even as it stands to change how we view meat, making meat more multiple, it rests on a narrow sense of how meat should be defined and consumed, when compared to the diversity of our past modes of carnivory. Cultured meat emerges at a moment when most of the meat that humans consume is like the hamburger itself, insofar as it is relatively homogeneous both in animal origin and in the forms in which we eat it.

Cultured meat is a part of the history of meat rather than a deviation from it. But if we zoomed out to view the full timeline of our species' carnivorous behavior, we could only explain late twentieth-century and early twenty-first-century meat—the meat on which cultured meat is based—by guessing that radical changes had taken place in the prior hundred years, as indeed they did. Those changes were both qualitative and quantitative, and they derived from industrialization and urbanization. They began in Great Britain and North America in the mid-nineteenth century, through the development of new forms of figurative and literal infrastructure ranging from animal husbandry methods to refrigerated train cars, and that infrastructure eventually globalized, in the process becoming ever more refined, transforming meat worldwide.[12] It is not for nothing that the *Economist* uses a "Big Mac Index," assessing the cost of a McDonald's hamburger sandwich in different parts of the world, in order to compare the purchasing power of currencies; hamburgers had become universal enough for the gesture to be meaningful by the time the index was introduced in 1986. From 1960 to 2010, global meat consumption more than doubled, and it multiplied many more times in some rapidly developing countries, like China.[13] But this was only the latest wave of meat's modernization, a process that has transformed almost everything about meat, from who eats it, to how much of it they eat, to what they think counts as meat.

Another way to pose the question is to look at the protagonists. When cultured meat is created, whose ideas about meat will go into the bioreactor, and whose ideas about meat will come out? During the time of my fieldwork, almost all the actors involved in creating and promoting cultured meat were Westerners, most of them were Europeans or North Americans, almost all of them were under sixty years of age, and the majority of those were under forty. These demographic details matter. They affect the types of animals these actors think of as meat sources, and the styles of meat to which these actors have been exposed. Older eaters, possibly born before the industrializing turn in meat production in the mid-twentieth century (which built upon nineteenth-century foundations, many of them established in the North American Midwest), would likely have encountered different kinds of meat as children. Similarly, and despite the globalization of what is often called the Western diet, non-Western eaters might have different ideas about the role of meat in their lives.[14] Thus, cultured meat as a project responds to a specific moment in the history of meat, a moment that happens to be unique from the standpoint of human history. With a few notable exceptions, the imaginative resources of the cultured meat movement have been framed and limited by the versions of meat found in the industrialized Western world. While this chapter ranges widely through the history of meat, it focuses on the history of meat in Europe and North America, because those locales developed the version of modern, postindustrial carnivory that is currently globalizing with vigor.

As it happens, the first known piece of cultured meat produced, and served to an audience, bore no resemblance to a burger, nor did it resemble any conventionally appetizing serving of meat. This was a "cutlet" of frog cells, served in March 2003 in Nantes, France, as part of *Disembodied Cuisine,* an artwork created and performed by the Australia-based artists Oron Catts and Ionat Zurr. The "cutlet" was a tissue of *Xenopus* cells, marinated overnight in calvados, and then fried in honey and garlic. Frog legs, well known as a French dish, are seldom considered edible anywhere else in the Western world. The delicacy is said to have originated when medieval French monks managed to get the Church to define frogs as "fish," enabling the monks to eat a little more animal protein at a time when the Church limited their consumption of terrestrial meat. Although Catts and Zurr's explicit goal was to raise questions about public attitudes toward biotechnology, their performance had the collateral effect of raising questions about how modern eaters define the limits of meat, and how those limits shift with geography and time. "As we were going to be eating the first ever tissue-engineered steak in France," Catts and Zurr have written, "we decided to use frogs as a comment on the disgust that many French people express towards engineered food, a disgust that parallels the reaction of some non-French people towards the idea of eating frogs' legs."[15] Their gamble was that the disgust one feels at eating the wrong animal might be akin to one's disgust at eating a product of advanced biotechnology. Perhaps, the insinuation ran, they are the same kind of disgust. Catts posted notices advertising the event at the booths of local frog merchants. After the event he said, "Four people spat it out. I was very pleased."

For those craving a physiologically precise definition of "meat," Harold McGee offers a good one in his influential book *On Food and Cooking*. As McGee says, meat is muscle. Muscle tissue consists of coordinated structures of cells, or fibers, each of which may be as thin as an individual human hair, and which are filled with individual fibrils.[16] Those fibrils are themselves made up of filaments of actin and myosin, proteins that slide past one another in the action of contraction, triggered by the nervous system. That contraction will decrease the length of the overall coordinated structure of muscle. Muscle fibers come in two types: white fibers, which help animals move quickly or suddenly, and red fibers, which help animals exert themselves over longer periods. Animals that move more quickly, like lagomorphs (rabbits, hares, and pikas), tend to have more white fibers. Animals that exert force continuously over a longer duration, like whales, tend to have more red muscle fibers where they need them. White fibers are fueled by glycogen (a form of glucose), stored in the fibers themselves, whereas red fibers are fueled by fat and contain a kind of biochemical machinery for turning fat into energy. This machinery includes cytochromes (compounds, important in metabolism and respiration, that consist of a heme molecule bonded to a protein) as well as myoglobin (a protein that binds oxygen and iron), which gives meat much of its color. Muscle fibers do not contain fat, but clusters of fat cells are often located between those fibers and the surrounding connective tissue. Lean meat is, notably, roughly 75 percent water, 20 percent protein, and 3–5 percent fat. The fat does much to create the flavor of meat. The connective tissue surrounding muscle (visible in meat as the silvery "sheet" on many cuts) has two main functions. First, it establishes the structure of a muscle, and second, it anchors that muscle to bone. The specific types of cells that muscle is made from naturally matter when it comes to eating. But so does structure. As McGee puts it, "The qualities of meat—its texture, color, and flavor—are determined to a large extent by the arrangement and relative proportions of the muscle fibers, connective tissue, and fat tissue."[17] In terms of texture, meat has a "grain," and "we usually carve meat across the grain, so that we can chew with the grain."

There are reasonable objections to this reduction of meat to muscle. After all, it is born from, and supports, a culturally specific distinction between "desirable" cuts of meat and the undesirable ones, which are discarded as offal. And this abstraction of meat to functional anatomy ignores others aspects of meat, such as the way that the grass an animal eats contributes to the flavor of its fat, and thus transforms the flavor of its meat. But McGee's definition is useful in the case of cultured meat, both because it describes the version of meat that the meat industry would like to produce in great quantities and because it fits the version of meat that scientists are attempting to produce in laboratories. As of this writing, the exacting nature of muscle's structure poses a formidable challenge to the scientists who hope to create cultured meat. While some forms of meat, such as hamburger or sausage, are ground up, and thus less reliant on structure for flavor and mouthfeel, eating steak means eating something that depends on its "grain." Certainly, more complex structures may soon be achieved. Cultured meat draws on techniques being continuously developed and improved in regenerative medicine, a field in which scientists attempt to grow specific functional tissues intended for human medical transplantation. Functional muscle structures have been created using in vitro techniques.[18] Needless to say, money flows vastly faster, and more abundantly, to medical research than it does to cultured meat research (a waterfall compared to a leaking kitchen faucet), but more sophisticated forms of cultured meat, such as steak, may eventually be the indirect result of regenerative medicine's progress.

Meat's physiological character has helped make cultured meat imaginable for medical tissue engineers, but meat's symbolisms are multiple. Historians, anthropologists, and other scholars have found gender in it, and patriarchy as well;[19] a symbol of human power and domination over nonhuman animals;[20] the result of a process by which natural resources are organized and extracted; a sign of modernization; a sign of affluence; the food of heroes.[21] Conversely, and as the anthropologist Josh Berson argues, meat can be associated with economic precariousness as well, because the cheapest forms of meat are often more readily available to the world's urban poor than is healthier food.[22] Think of the hamburger once more: edible in the car, on the go between jobs, or on the street itself. The association between hamburgers and mobility began at American hamburger stands and drive-ins during the affluent, postwar baby-boom years, but hamburgers have adapted to suit the needs of precarious life in grim economic times. While observers of meat have much to say about its relationship with affluence, the precise nature of that relationship has remained controversial, debated in particular by policy experts in Western Europe and North America. Modernization theorists and international development experts often give meat a central role in the "nutrition transition" that societies are expected to make worldwide.[23] As they become more prosperous, the inhabitants of the developing world are expected to buy and eat more and more meat. One technical term for this is "income elasticity," the idea that, for a given consumable, demand tends to rise with incomes. Although the idea that meat is income elastic proposes no foundational mechanism driving the desire for meat, it nevertheless harmonizes with the notion that it is natural, perhaps even instinctual, to want meat.

Despite the attractive economy of the "meat is muscle" definition, not all meat is the same. Specific cuts of meat can be political, as Deborah Gewertz and Frederick Errington have shown in their book *Cheap Meat,* a study of "flaps," or fatty sheep belly meat deemed unworthy by New Zealand and Australian consumers but happily eaten by Pacific Islanders. In Papua New Guinea, flaps are central to a vision of a desirable life for people who may be well aware that more affluent, white eaters have already rejected this meat.[24] In this South Pacific case, flaps serve a kind of symbolic duty, capturing the way meat flips between affluence and relative poverty, security and precariousness. Flaps also stand for the intertwining of race, economics, and diet.

And meat's political significance has manifested in other ways, especially at moments when urbanization and industrialization, or dramatic shifts in market economies toward liberalization, have compelled governments to regulate meat's production or distribution. In the mid-eighteenth century, Denis Diderot and Jean Le Rond d'Alembert's *Encyclopédie, ou Dictionnaire raisoné des sciences, des arts et des métiers* included the statement that "le viande de boucherie est la nourriture les plus ordinaries après le pain" (butchered meat is the most common food after bread), testifying to a sense that meat had become not only a common food but an expected one, and that there might be political consequences if meat became unobtainable—giving the French government a reason to ensure meat's availability and accessibility to individuals from all classes. In France, the United States, and elsewhere, government interest in guaranteeing meat's availability and accessibility ultimately ebbed, replaced by a government interest in guaranteeing that meat supplies were reasonably healthy, and by government subsidies, both for grain producers who produced feed for animals and for the meat industry itself, all of which helped keep meat costs low for consumers.[25]

A particularly crucial ingredient for imagining cultured meat, which appears in the 2013 promotional film played at Post's hamburger demonstration, is the idea that it is natural for humans to crave and to eat meat. That is, we may be omnivores, but we have a special affinity for carnivory, a "meat hunger" for which there is no comparable grain or vegetable or mushroom hunger.[26] This idea is often connected, either causally or simply by association, with the claim that meat was an indispensable part of the process by which hominins (the members of the evolutionary clade consisting of our own species and our immediate, now extinct ancestors) evolved into *Homo sapiens,* probably first moving from *H. habilis* to *H. erectus.* Thinking about meat thus sometimes means thinking in deep, evolutionary timescales. It readily shunts us into the kind of "timeless" register prized by early generations of anthropologists, those professional "deniers of coevalness" (as described by anthropologist Johannes Fabian),[27] for whom contemporary "primitive" peoples often represented the developmental past of Europeans.

The notion of a chronologically deep affinity for meat, often linked to hunting, swirls through popular culture in the developed world. In the first decades of the twenty-first century, a version of this claim has been prominently visible in the "paleo" diet, which stresses that we should eat as our *Homo sapiens* ancestors supposedly did during the Paleolithic—after the emergence of physiologically modern humans, in other words, but before the Neolithic Revolution, when the transition to agricultural settlement is thought to have begun ("Paleolithic" and "Neolithic" denote periods in a register defined by technological change). Most paleo diets involve consuming ample quantities of lean meat, along with fruits and vegetables, but little or no refined flour, sugar, or other industrial food products. Paleo promoters claim that it thwarts the disease patterns of contemporary civilization, including heart disease and cancer.[28]

That the paleo diet has been discredited by dieticians, anthropologists, paleoanthropologists, and others, that scientists have sneered at the effort to make us healthier by returning to an imagined hereditary past—none of this eliminates paleo from the popular culture.[29] And the paleo diet and cultured meat can, from the right angle, seem like each other's mirror images. Both are premised on the "sickness" of our modern, industrial food system. One gazes into the past and finds better meat there, thought to be a guarantor of human health for contemporary adults, dietary refugees fleeing flour and sugar. The other gazes into the future and likewise finds better meat, imagined as an aid toward environmental stability, the protection of nonhuman animals, and, yes, human health. The intellectual historian Arthur Lovejoy once spoke of the "metaphysical pathos" of an idea, a kind of charismatic chain of association conjured when that idea is mentioned, drawing the reader in. The idea that we might tap into our preagricultural condition and find the "optimal" diet indicated by our genetics certainly offers the metaphysical pathos of archaism, of living in harmony with our inherited bodies, and archaism is sometimes more alluring than futurism. It appears to offer certainty in place of risk. One of the distinctive features of the paleo diet is that it presents a hypothesized evolutionary past as a plan for our hypothetical dietary future, yoking archaism and futurism together.

The popular idea of meat's natural "fit" with our bodies has expert cousins, hypotheses developed by scientists in the fields of paleoanthropology and physical anthropology more broadly, and in primatology. To fully address the arguments that meat eating helped "make us human," in the long-running and complex speciation process that separated us from the rest of the largely herbivorous primate order, we would have to cover a challenging array of details. Just when did humans differentiate enough from our ancestors to earn the label *Homo sapiens*?[30] What kinds of evidence, including fossilized human remains, faunal assemblages, and primitive stone tools, do we have to work with? From when does that evidence date? When we say that "meat made us human," do we have in mind a speciation process of relatively short or very long duration? Lastly, and most annoyingly for those fond of essentialist claims about the human condition, just what does it mean to speak of "being human"? What physiological, cognitive, and social condition does that phrase imply? As our understanding of the human physical (and especially the genetic and epigenetic) condition increases, such statements seem less coherent. Do such statements only describe the "human" cells in our bodies, produced by an interplay of genetic heredity and environmental influences on gene expression, or do they also account for, say, the gut flora (among other floral assemblages) that constitute our bodies' microbiomes? Needless to say, such details are part of scientific literatures that are in a constant process of change and revision, as new findings are turned up at digs and new hypotheses are proposed and debated.

Some scientists have credited ancestral carnivory with helping to produce our modern physiological, cognitive, and social condition. Carnivory has been linked to our having smaller mouths and weaker jaws than our ancestors, but also to our facility for cooperation, although many meat-oriented discussions of humans' sociability are not only about meat eating in general but about the specific meat-procurement strategy of hunting, and especially the hunting of large, gregarious terrestrial mammals such as deer or aurochs (the ancestor animal from which the modern cow descends).[31] Other scientists have proposed that it was through the activity of post-hunt meat sharing that meat made our ancestors smarter, by enhancing their social intelligence.[32] The paleoanthropologists and evolutionary biologists who construct such arguments work from a challenging evidentiary basis, much of it around two million years old, roughly the time when our ancestors seem to have added meat to their diets, plausibly first by scavenging from animal carcasses. This is considerably before the emergence of *Homo sapiens,* thought to have taken place between two hundred thousand and three hundred and fifty thousand years ago, depending on what traits one uses as benchmarks, within the Paleolithic.[33] For those who may wish to track this against other familiar benchmark dates for the establishment of human civilization, the earliest known examples of writing, uncovered at digs in Sumer (in what is now Iraq), have been dated to some six thousand years ago.

But many of the distinct evolutionary claims made on behalf of meat come less from material evidence taken from early hominin encampments—from knapped stone tools, from animal bones marked by those tools, from faunal assemblages, from the information contained within hominin skeletons themselves—and more from modern human physiology. That is, meat has become an attractive answer for many scientists who ask how we got the bodies we have. We deviate from other primates in striking ways, both physiognomic (including adiposity in relation to muscle mass; we are fatter, and less muscular, than other primates) and in terms of life course. Compared to other primates, our span of life seems extended and its pace seems slow, both in the developmental stages and in post-reproductive adulthood.

Our brains are larger and more calorically expensive than those of other primates, and in 1995, Leslie C. Aiello and Peter Wheeler proffered the hypothesis that the development of our brains has been related to our diets, and the development of our digestive systems, in a very specific way.[34] Our large brains, they argued, required calories that might otherwise have been used by our gut tissue, which is also calorically costly to maintain; so-called "expensive" tissue takes a lot of energy, whether its tasks are cognitive or metabolic. Thus, we must have found a way to get the calories we need without the large guts necessary to process a large amount of food. Our surprisingly small guts suggest that our ancestors had access to sources of easily bioavailable calories, possibly including raw meat, and likely including cooked plant or animal foods.[35] It is crucial to note that Aiello and Wheeler's 1995 paper, with all its merits (though it has not gone unchallenged),[36] offered a hypothesis rather than a claim of fact, but it is often read as doing the latter. Richard Wrangham's more recent version of the story of human encephalization presumes that the use of fire for cooking made the calories in both plant and meat foods more bioavailable. Notably, Wrangham's brief but memorable contribution to the conversation around cultured meat deviates from his book *Catching Fire,* where he placed less emphasis on meat alone and more on cooked foods of all types, especially tubers and other underground storage organs.

High-energy foods, Wrangham's story runs, have stood in a circular relationship with encephalization for the genus *Homo.* These foods helped us develop bigger brains, and our growing brains in turn gave us improved physical and social skills, useful for acquiring more food. One surprising dimension of Wrangham's inclusion in the 2013 film is that when *Catching Fire* appeared, it was understood by some readers as an argument about cooked plant foods specifically, running *against* a prior consensus that meat, cooked or otherwise, had been the most important dietary driver in human evolution.[37] Either way, the circular relationship between improved diet and improving skills, implied by the "meat made us human" argument, bears a striking resemblance to a related claim, namely that we are in many senses a self-made species, or, as biologist and science studies scholar Donna Haraway glosses one paleoanthropological consensus, "Our bodies are the product of a tool-using adaptation that predates the genus *Homo.*" Dietary adaptability is, in a very loose sense, the ability to adopt new tools.[38]

It is plausible that a paleoanthropological consensus will take shape in support of the idea that carnivory "made us" human. However, it is equally plausible that none will, given the relative weakness and indeterminacy of the available evidence. Or the activity of meat eating may come to stand less as a crucial ingredient for the attainment of modern humanity, and more as a sign of our species' impressive dietary flexibility, adapting to the foods available in a given time and place. Evidence supports the idea that meat became part of the dietary repertoire of our genus around two million years ago, and there are good (albeit often inferential) reasons to think that meat was part of an "improved" diet, in the strictest sense of bioavailable calories, for those early eaters, possibly the immediate ancestors of *Homo erectus,* which emerged some 1.8 million years ago (*H. heidelbergensis* emerged some eight hundred thousand years ago, and *H. sapiens* about two hundred thousand to three hundred and fifty thousand years ago). A diversified and truly omnivorous diet, one that included foraged plant matter and dug-up tubers as well as meat (obtained from carcasses or through organized hunting), may have given our ancestors a greater chance to survive and procreate. Such a diet would, in turn, have supported migration out of Africa and to a wide range of geographic settings, from places where plant foods are plentiful to areas near the Arctic where, for much of the year, humans must subsist mostly on animal foods. Even across parts of Eurasia, vegetation would have been insufficient much of the year, and organized hunting would have helped humans survive.

Regardless of the state of the paleoanthropological consensus, it is worth asking why the reductive claim that meat helped make us human has proven so magnetic. One answer is the ease and utility of the claim. It makes meat into an explanatory "hinge" between a natural state and a cultural one, as if our species had emerged from the former and now lives mostly in the latter.[39] In the late twentieth century, a very different effort to understand the relationship between culture and nature traveled under the name "sociobiology" and continues to do so as of this writing, in the early twenty-first. The term "sociobiology" was made popular by the entomologist E. O. Wilson in his book *Sociobiology: The New Synthesis* (1975),[40] and sociobiological thinking occasioned great controversy between evolutionary biologists and social scientists, as well as between biologists: Wilson's fellow evolutionary biologists Stephen Jay Gould and Richard Lewontin were among his most prominent critics.

As Mary Midgely put it, sociobiology's "mild and minimal definition is the 'systematic study of the biological basis of all social behavior.'"[41] The "new synthesis" announced in Wilson's title was of the biological and social sciences. He proposed to unify their claims, with implications for everything from individual psychology to social organization—and expanding into the philosophical provinces as well, taking ethics "temporarily from the hands of the philosophers" and "biologiz[ing] it," in particular by seeking an evolutionary explanation for altruism, which emerged as the core theoretical problem for sociobiology as practiced by Wilson.[42] How could altruistic behavior, which seems to confer no survival or reproductive advantage, have evolved?

How could it have grown so widespread, becoming a kind of universal trait present in all human societies? This is a question worth keeping with us, because for many proponents of cultured meat, altruism toward nonhuman animals is at the heart of this emerging technology's appeal.

The anthropologist Marshall Sahlins, one of the first significant critics of Wilson's sociobiology, describes that doctrine's basic explanatory principle, which is applied to all behavior, as "the self-maximization of the individual genotype."[43] The fact that culture does not reduce easily to biological utility, and serves many ends beyond it, is one great theme of the critique of sociobiology from the precincts of cultural anthropology, both for Sahlins and for those who have written after him. Another is the fundamental difference between nature and culture, and the importance of insisting on that difference in the face of an admittedly impressive sociobiological alchemy through which nature and culture begin to resemble one another, frustrating our attempts to know and understand what is different and distinctive about either one. Sahlins's insistence (in 1976) on the separate dignity of culture is a complex matter. For one thing, it was an intellectual position taken in an ongoing turf war between physical and cultural anthropologists. For another, the stakes of relating culture and nature to each other, almost always to the advantage of one or the other, have long been political.[44] It is also complex because so many attacks have been made on the validity of a bold line between culture and nature, though the implications of erasing that line seem to vary with the political orientation of the person holding the eraser.[45]

Wilson's sociobiology was itself an impressive synthesis, woven out of work by many evolutionary biologists going back to Darwin. Still, a foreglimmer of the intellectual posture that underlay Wilson's 1975 book can be found in the political philosophies of early modern Europe.[46] In *Leviathan*, Thomas Hobbes established a circular relationship between claims about human social behavior and claims about nature, thus helping inaugurate a tradition of defining each in terms of the other. This eventually became, as Sahlins puts it, discovering "the lineaments of the larger society in the concepts of its biology." For Sahlins this discovery, still ongoing in sociobiology, is pernicious. It entails a category mistake for scientists of all stripes, but it is also a snare for our political thinking because it finds in human nature an "origin myth," explaining and justifying a variety of human social practices, including modern forms of market capitalism.[47] For example, market capitalism's origins could just as reasonably be found in a series of accumulating forms of exchange and enterprise, as in a root competitive instinct.[48] In the

1970s, Sahlins argued that sociobiology seemed poised to reduce modern culture to "a capitalist nature," with nature itself rendered in the colors of bourgeois capitalism.[49] This point is significant for the larger conversations about global problems and species-wide behaviors (including carnivory) to which cultured meat is connected. In this context, the idea of an evolutionary basis for meat-hunger has its own "metaphysical pathos" not so different from the pathos of antiquity that haloes the paleo diet. The idea that it is natural for us to crave meat, for evolutionary reasons, has the charisma of a kind of seamless knowledge that moves swiftly from theory to practice. As one analyst of sociobiology puts it, sociobiology often takes the form of "myth making" "in which scientific theory and fact are used as props for ideological and moral agendas," often through the construction of a different type of "prop"—an account of what humans are like, not genetically or behaviorally, but *essentially*.[50] Thus, the problem may be less the narrow technical claims of sociobiology itself and more their rendition, in the popular consciousness, as the message that biology equals destiny. "We polish an animal mirror to look for ourselves," writes Haraway, in a different context.[51] Indeed, we look at animals (including pre–*Homo sapiens* hominins) not just to understand ourselves, but to ground ourselves in what Haraway terms a "pre-rational, pre-cognitive, pre-cultural essence."[52] Cultured meat is also natured meat, meat that seeks legitimation through an account of our nature.

The lesson from the sociobiology debates is that even if meat helped "make us human," perhaps as part of a subsistence strategy based on diversified eating, this does not mean that it is in our natures, however nebulously or narrowly defined, to crave meat. Of course, meat need not be instinctually craved to be delicious. McGee suggests that even if we are not "wired" for meat (to use the metaphor of a self-consciously technological age), we do crave many of the nutrients meat contains, including long-chain fatty acids, essential salts, sugars, the iron bound to heme molecules, and vitamins A, E, and B12. While the cell walls of plants are thick, meat cell walls are fragile, a highly bioavailable source of nutrition waiting inside them. Meat, especially when cooked, provides calories and nutrients in concentrated abundance. From the strictest standpoint of individual nutrition and taste, there are good reasons to eat it, even to crave it, without recourse to a theoretical "essential" longing for meat. Instead, we could view meat less as a necessity for human life and more as a food for which humans have an elective biological affinity. Meat has been a part of a series of adaptive dietary strategies in which we, or the hominin ancestors that became us, have engaged for tens of millennia. It may well have played an important role in our evolution into *Homo sapiens,* but then, there is evidence that so did the use of tools and fire. To name any of these behaviors as definitive for the human condition is to make something other than a biological claim. It is to interpret a set of evolutionary selection pressures, or perhaps a happy match between nutrients and needs, as both essence and destiny. Centuries before the sociobiology debates, in his 1759 play *Candide: or, the Optimist,* Voltaire lampooned this way of seeing things: "Observe, for instance, the nose is formed for spectacles therefore we wear spectacles."

Paleoanthropologists have credited meat, and meat-acquisition strategies like hunting, with more than physical changes. Paleoanthropological narratives also credit meat with a role in human patterns of socialization. In fact, meat's status as a food with a "charismatic" social presence seems most obvious in cases in which hunting was a relatively peripheral subsistence activity for hunter-gatherer societies. If we can trust the evidence indicating that hunting mammals for food (as opposed to fishing) tends to be a less efficient way to garner calories than foraging, gathering, or gleaning, we may infer that hunting may have served purposes beyond simple nutritional provision, and had different social meanings altogether. Important here is a point first made by Sahlins in the mid-1960s, in an essay entitled "The Original Affluent Society," and which he further developed in the early 1970s. Rather than assuming that "primitive" foraging societies have frequently been on the brink of starvation, routinely saved by victorious hunters returning to camp, we should assume that hunting and gathering societies found that their subsistence strategies were effective given relatively low levels of time and energy invested.[53] And gathering was likely more efficient than hunting. Before detailing some of the precise changes in sociability that have been credited to hunting, it is important to note that Sahlins's portrayal of hunting and carnivory runs in stark contrast to the widespread modern idea of meat as a measure of sustenance. Meat has often been treated as a measure of wealth and plenty, but this has rarely been because it provided the bulk of the calories that sustain human life.

Hunting has been given a specific and influential role in developmental narratives involving meat. William Laughlin presented an unusually strong version of the claim in 1968: "Hunting is the master behavior pattern of the human species." He went on to say that hunting "involves commitments, correlates, and consequences spanning the entire biobehavioral continuum of the individual and of the entire species of which he is a member."[54] Not all of Laughlin's immediate contemporaries within physical anthropology agreed with him.[55] At the 1966 symposium where Laughlin first made his claims, a landmark event for the field entitled "Man the Hunter," there was a broad consensus that hunting was far less definitive for the archaic human condition than the symposium's title suggested. In particular, Richard B. Lee's contribution, "What Hunters Do for a Living," argued that the hunting of mammals provided only about 20 percent of the food for then contemporary hunter-gatherer societies. On this account, hunting is much less reliable, as a mode of subsistence, than gathering, yielding fewer calories per person-hour expended in the field away from camp. In one Bushman community Lee described, an hour of gathering could yield some two thousand calories, whereas an hour of hunting (averaged over the many hours a hunt often takes) yielded some eight hundred, and this despite the more concentrated calories contained in meat. In other words, hunters don't really hunt for a living. None of this, however, detracts from the social importance ascribed to hunting, including in such groups as the Hadza, in contemporary Tanzania, whose members may eat a diet consisting only of 20 percent meat, but who nevertheless say more about meat than vegetables when describing their food to outsiders. Certainly, meat, and the animals whose bodies yield meat, loom larger in the symbolic registers of many cultures than do vegetables, even when plant foods make up the bulk of the diet.

To say that meat helped make us human is one thing; to attribute our humanity to hunting is another. The latter defines humanity as a predator condition, and nonhuman animals as prey. In his *Meat: A Natural Symbol,* anthropologist Nick Fiddes argues that the most important function of meat in human life is not dietary at all. Rather, meat symbolizes our control over the natural world, our dominance over the rest of *Animalia,* our distance from the "lower orders." Dominance is also differentiation, a way of juxtaposing ourselves, along with that ineluctable human possession we call "culture," against a nonhuman nature. Many of our meat-eating practices derive their traits from that juxtaposition in a complex fashion, for meat (especially red meat) comes to simultaneously represent our power over nature and our animality—both our condition as animals and our will to escape that condition. To transform raw meat by cooking is to domesticate it, allowing us to consume meat browned instead of bloody. If the broader point of Fiddes's argument is to expose the uncertainty and ambiguity in modern meat eating, its dual conveyance of our mastery over nature and our animality (which is to say, our status as part of nature), the argument's additional implication is that the human condition is essentially predatory. Fiddes's argument shows a debt to the structuralist anthropology of Claude Lévi-Strauss, who saw significance in specific coordinations of nature and culture: cooking might transform the former into the latter, revealing the categories to be permeable, but the difference between them is crucial to the overall system of meaning that those categories create. But in its emphasis on the importance of predation, Fiddes's narrative does find support in one aspect of the anthropological literature on hunting: in many communities, hunting seems to play a more central role in social and cultural life than it does in nutritional life. In other words, from a nutritional standpoint meat's persistence as a symbol begins to look like overrepresentation—a phenomenon also visible in contemporary nutrition debates, which are, perhaps surprisingly, still shadowed by nineteenth-century claims about the human need for protein.[56] In the deep time of *Homo sapiens,* meat may have been as much a signal of status as a source of sustenance.

If meat's capacity to signal began with the hunt, it persisted through the use of domesticated animals as food sources.[57] The first domesticated animals were not for eating. Dogs were domesticated first, and they seem to have been domesticated some twenty-one thousand years before the first plants. This suggests that they were companion animals for hunters and gatherers.[58] Physical anthropologist Pat Shipman has suggested that animal domestication might usefully be thought of as "an extension of tool making," which led to domesticated animals that served as "living tools" and providers of "valuable natural resources."[59] Paleoanthropologists have long appreciated the fact that the advent of plant and animal agriculture did not lead to an immediate increase in human well-being. In fact, evidence suggests that early agriculture yielded fewer calories than prior strategies of subsistence via gathering and hunting, and that early generations of agriculturalists were physically smaller, and shorter-lived, than comparable gatherers and hunters. It is hard to definitively answer the question of why early humans engaged in agriculture at all, but some paleoanthropologists surmise that "intensification" strategies, for increasing the yield of plants from a particular patch of ground, became more attractive when climate change led to shortages of the plants available through non-intensive gathering practices. Other explanations are based on the fact that agriculture makes more sense the larger a population it feeds. Hunting and gathering can support smaller groups of humans, but larger groups quickly exhaust the resources around them.

Regardless of its bad effects on early generations of adopters, the agricultural adaptation eventually became the dominant pattern for *Homo sapiens* in many parts of the world, facilitating larger and denser human settlements. The domestication of food animals became a central fact of human life. Some have even pointed to the physical shifts in humans caused by animal domestication; Shipman describes this as reciprocal domestication, one of whose hallmarks is the functioning of lactase in adulthood, enabling many human populations to consume dairy products safely, past childhood.[60] While humans may have first attained a high trophic level (i.e., living near or at the top of a food chain) by eating animals as scavengers and hunters, it was the domestication of animals that would ultimately secure their position.

A complete map of meat's history would have to include the myriad transformations of animal husbandry and agriculture through which meat ceased to be solely the product of scavenging or the hunt and became a predictably available food. Short of such an exhaustive but unwieldy map, it is worth noting the most striking differences between the meat we eat today and the meat consumed just a few hundred years ago (in Europe), or more recently than that in other parts of the world. McGee calls this the dominance of the "urban" over the "rural" style of raising, killing, and eating animals.[61] During the last few hundred years of urbanization and industrialization, the rural style has been all but lost. Roughly speaking, the rural style involves a long period of human coexistence with animals prior to slaughter and, often, the use of animals as workers before they are killed for food. Consumed at an older age, such animals tend to have a stronger or more developed flavor. By contrast, the meat of the younger animals consumed according to the practices of the urban style tends to be more tender, less lean, and milder. This is the meat to which many twentieth- and twenty-first-century eaters, especially in the developed world, have grown accustomed, though in the late twentieth century, and especially in the United States, some consumers began to prefer leaner cuts from the same animals. In 1927 the urban style of beef received government support in the United States through the Department of Agriculture's introduction of a grading system, based in part on fat marbling; this grading system was informed by a series of public hearings, conducted across the country in 1925, designed to give the meat industry a voice.[62]

The dominance of the urban style is not due to industrialization alone. It is based on a long series of processes through which animal bodies and technology were coordinated together. A set of infrastructures were created, including everything from the supply chains that bring meat to consumers, to breeding systems for producing new animals, to more abstract kinds of infrastructure such as the markets for sought-after products like the semen of prize breeding bulls. From the standpoint of the cultured meat movement, these infrastructures are distinctively wasteful, environmentally damaging, and cruel. In the form of CAFOs, they also present deadly breeding grounds for zoonotic pathogens, even as the subtherapeutic doses of antibiotics given to animals help build antibiotic resistance in those same pathogens. But these forms of infrastructure are also necessary if billions of people are to consume conventional meat in the volumes expected by global eaters on the Western diet.

The mid-twentieth-century architectural historian and critic Siegfried Giedion drew a line between what he called "living substance" and "mechanization." In his 1948 book *Mechanization Takes Command: A Contribution to Anonymous History,* he sought to chart the effects of modern technology on both human and animal bodies, in part through an analysis of stress: the wear on joints, muscles, and soft tissue of industrial labor, and the effort to make animal bodies fit into standardized pens and assembly lines and to make them perform as if they themselves were machines. Here it is crucial to note that industrialized meat production was not a system of mechanical harms inflicted on animals that were the direct equivalent of their wild ancestors, in behavior and in physiology. Industrialization effectively accelerated or intensified the productivity of an agricultural system that was already technological in many senses, and whose parts included animals that were bred for generations to suit human needs. This is not to say that these living parts did not suffer. Their bodies were the organic products of breeding systems that could never bring livestock perfectly in line with the processes of slaughter, butchery, transport, and sale.

And yet, even if animal agriculture became a technological system long before industrialization, it seems clear that the new methods of breeding, raising, managing, and killing animals, especially in the twentieth century, applied an additional, and qualitatively different, pressure to animal bodies.[63] These methods produced not only more meat, but also unforeseen consequences, including health risks for both human and nonhuman animals and high levels of industrial pollution. These new ways of treating animals included a better understanding of the genetics underlying desirable traits, and of animal nutrition and health up to the desired age and weight of slaughter (perhaps most especially via the use of antibiotics), as well as improvements in feedlot and abattoir design. All this rested on knowledge gleaned at agricultural colleges and experimental stations, which the law and natural resources scholar William Boyd has described as a shift from animal husbandry to animal science.[64]

As mentioned at the start of this chapter, modern meat consumption in the Western world is characterized by high volume coupled with a narrow range of types of meat eaten; we have grown dependent on a few species particularly amenable to domestication, especially cows, pigs, and chickens. To account for the increase in volume, economic historians have often employed the idea of meat's income elasticity. As Friedrich Engels wrote in *The Condition of the Working-Class in England in 1844:* "The better-paid workers, especially those in whose families every member is able to earn something, have good food as long as this state of things lasts; meat daily, and bacon and cheese for supper."[65] Engels went on, telling his version of a story that we see elsewhere among social scientists, namely the story that demand for meat is income elastic, rising and falling depending on available monies: "Where wages are less, meat is used only two or three times a week, and the proportion of bread and potatoes increases. Descending gradually, we find the animal food reduced to a small piece of bacon cut up with the potatoes; lower still, even this disappears." Meat's status as a special luxury, sometimes unavailable, changed remarkably within decades of Engels's observation, as the declining price of industrial meat made it a staple food for the masses. When Engels did his research, the phenomenon of cheap meat, making previously luxurious cuts of meat available to nearly everyone, had not yet arisen.

As late as the late eighteenth century, meat was a rarity for most people in many parts of Europe, perhaps consumed at Easter or other holidays, or at times of unexpected plenty. Only those in particularly privileged positions consumed more, and "cheap meat" as we know it did not exist. While meat consumption rose considerably in Europe during the nineteenth and twentieth centuries (it was, in fact, on the rise in Engels's), in other parts of the world such increases came later. Europeans in China, in the early twentieth century, noted that in some parts of the country (such as the north) most peasants ate meat only a few times a year.[66] A critical condition for the rise of meat consumption was the increasing availability of crop plants for fodder, made possible by innovations in nitrogen fixation to produce chemical fertilizers, by increasingly industrial and mechanical agricultural practices, and by the use of higher-yielding cultivars.[67] A host of technological innovations made way for the "high-volume thrift"[68] of industrial-scale agriculture.

Meat has become cheap in more than an economic sense. It has become experientially cheap as we live farther from the cycle of animal husbandry, slaughter, and butchery. In fact we are often completely isolated from it, in part because the meat industry takes measures, such as forbidding the use of recording devices inside feedlots and slaughterhouses, to ensure that it can control how industrial animal husbandry is represented in the media. Billboard advertisements for meat, as well as supermarket packages, often avoid directly representing the animal from which the meat has been taken.[69] This issue of packaging reflects a marked change, one that took place in the course of the later nineteenth and twentieth centuries. As William Cronon points out, in the city of Chicago, the nerve center of the U.S. meatpacking industry, attitudes toward meat shifted in the late nineteenth century:

> Formerly, a person could not easily have forgotten that pork and beef were the creation of an intricate, symbiotic partnership between animals and human beings.... In the packer's world, it was easy not to remember that eating was a moral act inextricably bound to killing.... Meat was a neatly wrapped package one bought at the market. Nature did not have much to do with it.[70]

The first animals that moved through Chicago's packing yards had been pastured in America's "Great West," and they enjoyed lives greatly preferable to those of their counterparts a hundred years later, which would conform more to McGee's urban style of meat—killed young and tending to live more of their lives in an entirely industrial context. But the mechanisms of experiential distancing had set in earlier, and they would make the shift from a rural to an urban style of meat consumption all but invisible for many American consumers.

In considering the relationship between cultured meat and the history of meat, it is crucial to remember that the style and volume of our modern carnivory has been possible for only about a hundred years. It is just one of a cluster of startling transitions modern foodways have seen. Rachel Laudan has coined the term "middling cuisine" to describe a broad leveling of food practices in modern Europe and the emergence of a cuisine that was "middle" not in terms of its quality but in its polyglot character. Middling cuisine incorporates elements from both high and low cuisines, as well as foods from other parts of the world. Thus, curry powder made its way from colonial India into middle-class English bowls. Differences between the food eaten by the rich and the poor did not disappear, but class ceased to be an insuperable barrier when it came to foodways within a single culture. Middling cuisine,

Laudan writes, "ran in close parallel with the extension of the vote."[71] And with the rise of middling cuisine, meat consumption increased, as did the consumption of fat and sugar (no longer used only as a medicine or spice), all of which is collectively referred to as modernity's "nutrition transition" and is blamed for many of the chronic diseases that plague modern eaters in the developed world. Much as the gradually decreasing cost of sugar put the sweetener in everyone's tea,[72] cheap meat made a democratically carnivorous middling cuisine possible. But if carnivory at small scales poses few problems to human health or to the natural environment, this same claim cannot be made for industrial animal agriculture. This is the dark side of the meat industry's ability to produce meat that billions of people can afford to eat regularly.

From the standpoint of the long history of human carnivory, industrial animal agriculture, as practiced in the late twentieth and early twenty-first centuries, has been a stunning departure from prior trends. It is almost as stunning as the rising global human population, which had reached about 1.2 billion by the time meat began to truly industrialize in the mid-nineteenth century; as of this writing, it is estimated at 7.5 billion. The term "modernization" glosses such radical changes, which also include urbanization and the aforementioned rise of middling cuisine in much of the world, almost too easily. The world became modern in so many transformative ways that it became necessary to ask if different aspects of modernization might be linked, with urbanization, population growth, and dietary change causing or encouraging each other in ways that social scientists continue to puzzle through. There are more trend lines, to be sure. Some watchers believe that meat consumption in the developed world is already beginning to decline, at least within certain demographics.[73] There are reasons to believe that meat's cost will rise, along with other food costs, as the world loses farmland to climate change, forcing consumers to make hard decisions about how much meat to continue eating. And some nutritionists expect a near future in which we meet our protein needs by tapping an increasingly broad range of sources, from familiar ones like legumes to ones that Westerners tend not to eat, like grasshoppers—all of which would entail a return to a more capacious definition of "meat." Thus, the question that began this chapter remains: if cultured meat does emerge and transforms the food system, will it be an effort to reproduce the industrial meat-forms we know, albeit on a novel, and more ethical and sustainable, foundation? It is too simple to say that cultured meat promises to satisfy the human appetite for meat. Cultured meat's creators might also preoccupy themselves with the question of which human appetite for meat, in historical terms, they wish to satisfy. Meat is much more than food. Meat may have helped make us; meat may help kill us. The strangest thing about cultured meat is its very assertion of meatiness: cells cultured in a lab are called meat, buttressed by the thought that meat was always like this, and that we, as eaters, were always as we are now. This is a pitch about the future that believes it also needs the past.

THREE

Promise

Fast forward two years. The twin of Mark Post's 2013 hamburger lies on a white plate. Plastinated for preservation, this is the other patty Post and his team handcrafted in the weeks leading up to the cultured beef demonstration in London. The resulting object looks like a wan hockey puck to my North American eyes, and it is boxed on its plate under Plexiglas at the Boerhaave Museum of the History of Science and Medicine in Leiden, Netherlands. Post has donated it to the museum's permanent collection, but on this particular day it is part of a larger exhibit on the future of food. Somewhat distractingly, it shares its pedestal with a device for extracting semen from bulls, an artifact of twentieth-century livestock breeding. Like cultured meat, this is a technology for multiplying beef, albeit through in vivo rather than in vitro methods.[1]

The burger has been enshrined as part of the historical record before its effects on the food system (if there will be any) play out. I didn't see the burger until 2015, almost two years into my research, but it feels right to mention it now. Through the voice of Hamlet, who says, "Time is out of joint/oh cursed spite/that ever I was born to set it right," Shakespeare compares time itself to a finger, in a grand metaphor for unease. And in a late twentieth-century cinematic rendering of meat as an image for time, a history professor announces to his class, "Never forget, my father was a butcher," and from his briefcase he takes a cutting board, a metronome, and a cleaver, followed by link after link of blood sausage. A student volunteer cuts the sausage in time to the metronome—there are no joints for time to be out of here. "Marx thought Man would one day stop eating sausage," the professor says. "Einstein ripped off the skin, and then it lost its form." This is a scene in Alain Tanner and John Berger's 1976 film *Jonah Who Will Be 25 in the Year*

2000, a meditation on the legacy of 1968 when, for a generation of young Europeans, time seemed to lose its old shape and gain a new one. "I'm going to talk about how the folds of time are made," says the professor, referring either to the workings of power on the substance of history or to "the events" *(les événements)* that transformed cultural and political consciousness one European summer. What kind of refolded time has cultured meat created, what kind of departure from "jointed" time?

There is a reason I'm getting ahead of myself. Expectation pulls me. The psychoanalyst Jacques Lacan once wrote of "the assertion of anticipated certainty," which can make time seem to speed up or warp ahead: "the haste function," he called it. The burger display at Boerhaave implies that the burger will be so important that it already belongs in the historical record. As the literary critic Shoshana Felman has pointed out, such assertions of certainty often take a very familiar verbal form, namely the promise.[2] Promises are usually expressed in speech or writing, but objects can be promises too. Here are a few haste-inducing statements made on behalf of cultured meat as of 2013:

We proved it's possible.[3]

In vitro technology will spell the end of lorries full of cows and chickens, abattoirs and factory farming. . . . It will reduce carbon emissions, conserve water and make the food supply safer.[4]

If we shifted to cultured meat . . . it would reduce greenhouse gas emissions more than everybody trading in their cars and trucks for bicycles.[5]

The intellectually responsible reader will be quick to point out that these statements are not formal promises. The first statement, made by Post at his burger event, was meant literally. Of all the people working on cultured meat, Post is among the most levelheaded and invested in transparency about possible outcomes. The third statement is an "if-then," offered by Jason Matheny, the founder of the nonprofit organization New Harvest, in 2010. It conveys confident knowledge about the benefits a new technology could yield but says nothing about the precise causal relations that would bring those benefits about. Only the second statement, with its dual "wills," comes close to a promise, though it is still less than a contract written up for the reader to countersign. It comes from PETA, an organization known for bold statements in support of protecting animals. Yet all three statements can, in the right context and for the right listener or reader, seem promissory. This prompts questions. If the maximal form of a promise might be a contract (in a legal culture) or a prophecy (in a religious one), what's the minimal form of a promise? How casual can a statement about possibility be, and still create a sense that something has been promised? The answer to this question will vary case by case, with the moral weight of what is at stake and with the eagerness of audiences. There is a world of difference between promising a treatment for degenerative blindness and promising a genetically modified apple whose exposed flesh does not brown in the air.[6]

This leads us to the question of how promises affect us. Felman and Lacan are keen to answer, seeing the promise as a kind of "speech-act" whose goal has less to do with truth than with action, with moving from language to deed. Promises are unstable, for we never really know if we can keep them. We use these makeshift verbal tools to make our imagined futures a little less unsteady. Anyone who says there will be cultured meat in year X knows that there might not be. However, even if an entrepreneur, scientist, or pundit appreciates contingency and the possibility of shipwreck, they may not be willing to talk in public about such things, which might seem to betray an essentially promising technology. This sort of futurism is a set of performances, a matter of building confidence, which means it has a lot in common with confidence games. Over the years I followed the unfolding story of cultured meat, public discourse around it gradually grew less free-spirited and open, particularly when it came to the question of this substance's technical and commercial viability and eventual benefits. This, I believe, is due to the hype and spin that inflects most, if not all, conversations about emerging technologies in which venture capitalists have taken an interest. When one entrepreneur expresses great confidence, it raises the confidence stakes for everyone else.

It is worth pausing over Friedrich Nietzsche's passage on promises and promising in his *Genealogy of Morals:* "To breed an animal with the right to make promises—is not this the paradoxical problem nature has set itself with regard to man? And is it not man's true problem?"[7] There are puzzles here that can temporarily negate the effects of the "haste function" and let us breathe and think. Nietzsche compares humans to animals bred to a purpose, as if we were Nature's domesticates. Nature is implicitly anthropomorphized even as the human loses a crucial degree of difference from farmed animals. But even as he rubs out the chalk line that divides humans from their fellow creatures, Nietzsche also redraws it. He suggests that in breeding us, nature created an animal that has the "right" (a curiously legal term in this context) to make promises. In other words, humans are distinguished from all the other animals not by our brains, not by our relative hairlessness, nor by our use of fire, but by our promising. Felman reads this passage and notes that in establishing us as the promising animal, Nietzsche offers an alternative to a much earlier philosophical characterization of humans. Aristotle called "Man" "the political animal,"[8] not only because of our sociability, not only because we are born into social structures upon which we become dependent, but because our power of speech allows us to promise.

Mike Fortun, reading Nietzsche's passage, notes that promising's very nature involves paradox. A colloquial expression of doubt, "promises, promises," shows that a mere doubling of the word inverts it and dumps all of our trust on the ground. On some gut level, we know that promises are merely hopeful ways of projecting certainty beyond certainty's native temporal range. Promises extend the will beyond its reach and grasp, beyond what it can control. If promising helps bring a particular future about, it will seem to have retroactively justified our belief in the original promise. Another way to say this is that promises, particularly when they are personal, can recursively bind their makers and their receivers over time. Promises are, Fortun writes, "language run[ing] on credit, if you like—which is not to say, as you surely know, that bills don't fall due."[9] The paradoxical character of promising matters because promises are an inevitable feature of human life. We have to treat them as if they are trustworthy, though we know they are inherently un-.

As Hannah Arendt put it when she interpreted Nietzsche's passage in her 1958 masterwork *The Human Condition,* promising relies on the faculty of the will in order to raise an island of certainty in an impossible-to-know future. Arendt, who noted that Nietzsche was preoccupied with the individual will, focused her attention instead on promising as a shared, collective, and ineluctably social activity, one of the best examples of which is political sovereignty itself. Sovereignty is produced by collective and concerted action, which is in turn made possible by our capacity for "mutual promise or contract."[10] Sovereignty thus "resides" in what promising produces, which is a "limited independence from the incalculability of the future." The limits of this independence from incalculability are "the same as those inherent in the faculty itself of making and keeping promises."[11] What a properly sovereign people gains is the ability to "dispose of the future as though it were the present."[12] This kind of sovereignty is only possible for a collectively promising political community, however, and not for a people who receive, more or less passively, the singular promises of a singular leader.[13]

In all areas of life in which we hope to "dispose of the future as though it were the present," much depends on the promise-makers. Many intimate promises, such as those made in marriage, are kept up by the "credit" of our belief in one another. Technological promises are conveniently different, because there is an available source of impersonal credit to underwrite them. This is the common belief in technological progress as a driving force in history, which historians sometimes formalize with the term "technological determinism."[14] This belief has its variants. Sometimes it means believing that inventions have autonomous lives of their own, perhaps changing the world through their intrinsic characteristics rather than their patterns of use—from the development of calculating machines to a networked world of information, to offer one cribbed version of this way of seeing the world. In even more extreme cases, exemplified by some contemporary purveyors of corporate futurism, this means believing that technological progress has been the central protagonist of human history itself, from the knapped stone blade to biotechnology. It is a convention among historians to trace technological determinism back to the European Enlightenment, and to note that it has traveled hand-in-hand with the idea of progress.[15]

Potential is the ground of promising. Cultured meat relies on a discrete biological entity, namely the stem cell, which has been seen by scientists as full of potential, and especially medical potential.[16] In fact, cultured meat is, for all the fanfare that surrounds it, part of the larger story of the attempt to use stem cells for regenerative medical applications. The potential of stem cells to produce multiple somatic cell types, to effect healing, to create growth, is often described by scientists as a latency unlocked by laboratory technique.[17] Put differently, nature has a potential that only human culture can set free. The perceived potential (a word that derives from the Latin for "power," *potentia*) for stem cell therapies shades off quickly into a sense that those therapies are "promising." Informal networks of patients, or potential patients, grow in response to potential therapies, presumably in the hopeful grip of the haste function. Call them "communities of promise."[18] They await treatments or cures for diseases, such as Alzheimer's, and they feel keenly the gap between the actual and the possible. Often they do not merely wait, but advocate for funding to support research for their desired treatments. Sometimes informal networks become formal lobbying groups. In an odd way, the "community of promise" around cultured meat echoes the medical patient networks. Both have grown familiar with frustratingly long waiting periods, with the pace of bench research, and with the need to keep up morale while waiting.

I knew when I began my research, in the fall of 2013, that speculation in biotechnology involves participating in promises, whether one is an entrepreneur, a professional futurist, or a moral supporter of an emerging technology. Nor was I exempt. The grant that made my research possible, issued by the National Science Foundation, was given a title by that organization, "Tissue Engineering and Sustainable Protein Development," as if I were not merely an observer but also an agent of cultured meat work. I was going to have to get used to the promissory, because I was implicated in it myself. I mused over an etymological meditation from the twentieth-century French philosopher of biology Georges Canguilhem. He noted that the word "tissue" comes from the Latin verb *tisser* (to weave).[19] "Tissue," he wrote, was a term "charged with extra-theoretical implications," much as the term "cell" often conveyed historically contextual ideas about the relationship between parts and wholes, both in organisms and in social orders. If cells seemed to Canguilhem to always refer to natural models, weaving seemed firmly on the side of culture. "A cell makes us think of the bee and not of man," Canguilhem went on. Weaving makes us think of man and not of the spider: woven fabric "is human work par excellence." If cells are closed upon themselves, weaving is continuous, and interruptions in the woven fabric are arbitrary rather than inherent in the form itself. One can stop and restart at any time. Tissue engineering would then represent the potential endlessness of "woven" biological matter, and "endless" is a far more existentially loaded term than "sustainable." "One folds and unfolds a tissue," wrote Canguilhem. "One unrolls it into waves atop one another on the merchant's counter."

I hit walls in those early days. There were phone calls to make and electronic mail to send to every researcher in the field I could find, including Mark Post. The National Science Foundation's support, and the postdoctoral position at the Massachusetts Institute of Technology that it funded, gave me a kind of credibility. Still, conversations and invitations to laboratories were few and far between. Most messages were not returned. Frustration was inevitable. I was lucky to find an interlocutor in Isha Datar, the young biotechnologist who had recently become the executive director of New Harvest, a nonprofit organization that was founded in 2004 to encourage research into cultured meat. Over the months and years that followed, Datar would become one of my most helpful conversation partners, sharing her thoughts and views on cultured meat and the shifting dynamics of debate, funding, and enterprise around it. Prior to joining New Harvest, Datar had coauthored one of the most widely circulated early papers on cultured meat.[20] She would build New Harvest into an organization that, while functioning as a "community of promise" that connects celebrants of cultured meat, also encourages careful reflection on the environmental and ethical dilemmas that make cultured meat appealing. In between conversations with Datar and interviews with anyone who could spare a few minutes, I did a great deal of waiting and reading. I also visited other sites of speculation into food's possible futures, including consulting firms, think tanks, environmental nonprofits, and many technology conferences.

Datar, who is Canadian, moved from Toronto to New York, and New Harvest went from a desk in her apartment to an office suite, and eventually acquired more staff under her guidance. They also acquired start-up-like branding, including a logo reminiscent of cells growing in culture. New Harvest broadened its portfolio, creating the term "cellular agriculture" to refer to a range of animal products derived from in vitro rather than in vivo sources. The term is felicitous: the vocabulary of tissue culture research is already heavily inflected by agricultural terms like "growing" and "harvesting."[21] I also spent time with a start-up that hopes to create milk without cows, the founding of which Datar coordinated. Originally called Muufri, they would later rename themselves Perfect Day. Other start-ups targeted egg proteins or rhino horn, the latter with the goal of flooding the Chinese market for "medicinal" powdered rhino horn and thereby reducing illegal poaching.[22]

As I approached working laboratories, I hit more walls. Venture capital backing usually means the obligation to build and protect intellectual property, and this meant that I would often get a few minutes with a company founder or spokesperson, but no access to laboratories or to the valuable time of scientists. I learned that I was not the only one frustrated by the resulting cloud of unknowing; many cultured meat scientists simply cannot report on the state of the art at other labs and are thus at risk of reinventing the wheel. New Harvest has made an express commitment to supporting open-access research, which has mitigated this problem, but only to a degree. Meanwhile, the promises continued to appear and I sometimes worried about the dominance of the promissory, that the promise had become the only speech around cultured meat that mattered. All other kinds of speech, including debate about what is desirable in our food system, tended to fall by the wayside.

Back at the Boerhaave Museum, I am still in the time signature of Lacan's "haste function," thinking about the ways in which time goes out of joint in anthropological writing, which has traditionally reached out of the contemporary and into the human past in search of lessons about both nature and culture; it is only relatively recently that anthropologists have attempted to visit the future, figuratively speaking, by writing about the ways we anticipate it and try to build it.[23] Anthropologists have long imagined themselves as doing a kind of time travel through the simple act of journeying from a less developed part of the world, where they found their field sites, back to the more developed place from which they originally hailed. Home from the field, they have written as if their ethnographies took place in a time other than their own—a time imbued with a strange permanency—and as though their anthropological experience dropped into a bucket in a premodern world, to be washed and sorted in modernity. The anthropologist Johannes Fabian calls this "the denial of coevalness," the belief that contemporary "primitive" peoples are analogous to human populations of the distant past.[24] Trying to write anthropologically about cultured meat means experiencing the denial of coevalness from another direction. The researcher becomes the one living in the past, the one who hasn't caught up with the subject or topic of research. It's the reverse of the familiar experience one can have by strolling through an ethnographic museum at a university, especially one in a country that was once a colonial power (Pitt Rivers at Oxford University is one example; the Museum of Archaeology and Anthropology at Cambridge is another). Perhaps a totem pole from a First Nations tribe in the American Pacific Northwest is displayed there, as if the life of the tribe in question is still ongoing, as if modernization has not yet reached them, as if they are a "cool" culture and not a "hot" one.[25] But at Boerhaave, from the future standpoint represented by the burger in its Plexiglas case, I am the one out of joint, still eating animals.

FOUR

Fog

The bus I'm catching finally emerges from the fog. I'm in San Francisco. After months of making phone calls and sending emails, I am finally in the field, having gotten a lead. Above me the fog cancels the parrots as it cancels everything else, and then they burst down from the power lines where they had been perching. Red, blue, and green, the birds refuse cancellation. There are words written on the side of the bus: "Shareable Content," emblazoned across a McDonald's advertisement for Chicken McNuggets, which sit in their paper basket, paired with a little paper cup of dipping sauce.[1] In his short book of 1905, *Jokes and Their Relation to the Unconscious,* Sigmund Freud argues that most kinds of jokes function through a surprising juxtaposition of seemingly disparate concepts, or by forcing the listener to recognize an absurdity they have unconsciously noticed but have not brought into conscious awareness. Just so, the joke in this advertisement has its layers. The most superficial level is obvious: Internet terminology, like the marketing term "shareable content," invades our lives at every level. Baskets of nuggets are apparently for sharing with our friends in the same way denizens of the Internet share baby goats playing on seesaws—that is, in videos of them, sometimes marked by narrative reversal, the big goat surprisingly bumped off one end by the little goat, who jumps on his end with greater force. Less obviously, the advertisement seems to acknowledge that the nuggets' food value is similar to the value of much of what we share on the Internet. Junk food is light on consequences until we tally its health effects.

Freud contended that jokes produce pleasure by removing unnoticed but persistent psychic censorship. Unawares, we expend energy to censor our thoughts, often influenced by social cues about what constitutes appropriate thinking. Freud wrote that the "yield of pleasure" in telling or responding to a joke "corresponds to the psychical expenditure that is saved."[2] The visual comparison of chicken nuggets—which one study showed to be mostly composed of fat, plus fragments of bone, nerves, connective tissue, and epithelium (the tissue at the outer layer of the body), as opposed to lean white chicken meat[3]—with "shareable content" is subversive, and perhaps goes beyond the intentions of the firm that made the ad. It is almost as though the joke saves psychic expense by making fun of everything at once. One cultured meat researcher I chat with suggests that there could be some 875 million skeletal muscle cells in a chicken nugget, assuming that the nugget in question is actually made of muscle meat. By contrast, a hamburger might contain some forty billion cells.

I'm in town to visit one of the places where money has started to trickle in to cultured meat research, hoping to learn what cultured meat means to some of its benefactors. To the best of my knowledge, no cultured meat has been made here, but as of 2013, San Francisco and nearby Silicon Valley are the unofficial capitals of futurism in the early twenty-first century, places where people come to build, or to debate, futures. San Francisco is currently enjoying another gold rush, full of excitement about the future, and I'm reminded that financialization—the movement of capital from industrial applications to investment—has itself been described as a kind of gamble on the future, and sometimes even on the eventual arrival of a new industrial revolution.[4] But the city is also full of agitation, because in recent years San Francisco has become a national and global symbol for income inequality and the gulf between the winners in the technology sector and the vast majority of their neighbors. I used to live in nearby Oakland, across the San Francisco Bay, and I have cause to feel wistful on return visits like this one. I am not visiting the organization that manages Sergey Brin's charitable giving or contributions, such as the one he made to support Mark Post's 2013 hamburger project. Brin's people have, for complicated reasons, been impossible to reach. Instead, I have arranged to visit the offices of Breakout Labs, a branch of the Thiel Foundation, the philanthropic institution of businessman Peter Thiel.

As of 2013, Thiel is one of the most prominent investors in Silicon Valley, as well as one of the Valley's most controversial voices on the subject of the future.[5] Through Breakout Labs, he has also become an indirect benefactor of cultured meat research. Rather than making traditional investments, Breakout Labs gives small grants to companies operating in areas of technology development deemed especially likely to "turn wild ideas into world-changing technologies,"[6] as Breakout Labs' website puts it. In 2012, the year they were founded, Breakout Labs gave a grant to Modern Meadow, a company based in Mountain View, south of San Francisco, which as of 2013 hopes to use tissue-engineering techniques to create food products, as well as a leather-like "bio-material" intended for the fashion industry.[7] This stands out for several reasons, and not least because Breakout Labs is an unusual organization, since it gives philanthropic money to for-profit entities, namely companies involved in technology development, but what has my attention is its investment in a potential food. Food, after all, is a carefully regulated thing for reasons of public health, and Thiel is an outspoken libertarian. In a 2011 profile of Thiel for the *New Yorker,* George Packer suggested that PayPal, the online payment system that was Thiel's first big financial success and fortune-maker (along with his 7 percent ownership stake in a very prominent social media company), had its roots in Thiel's desire for "an online currency that could circumvent government control."[8]

I silently christen my bus "Shareable Content" and climb on board. The very first thing that happens, as we turn a corner to head north from Potrero Hill through San Francisco's comparatively small urban core, is that a tech bus de-fogs coming the opposite way. Although it isn't marked with the name of any of the big technology companies in the South Bay, there is no mistaking it—white and two-tiered, with an unoccupied bike rack folded against its rear, this is a tech bus. It is about 10 A.M. I've timed my visit so that my bus won't hit too much rush-hour traffic (though in an overpopulated city like San Francisco, rush hour is a permanent state of affairs). The approaching tech bus has plausibly just come back from delivering San Francisco–based workers to one of the campuses in Silicon Valley, or perhaps it is here to pick up a later shift. I've never worked on those campuses and I don't know for sure. I do know that Potrero Hill is one of the many neighborhoods of this city served by a network of buses, each run by subcontractors, who are in turn employed by the big tech companies to create a form of private mass transit, superimposed on the city's existing urban grid (in many cases the buses use public bus stops, for a nominal fee paid to the city) and intersecting in various ways with existing, public mass transit.[9] The buses are symbolic of the city's most recent transformation, and in the eyes of some locals this transformation is so undesirable that piñatas in the shape of tech buses are broken with baseball bats at protests, rallies, and parties.[10]

A generous friend has put me up at her Potrero Hill apartment while I'm in town to do fieldwork. On her refrigerator door is a poem cut from the pages of the *London Review of Books,* "Hollyhocks in the Fog," by the San

Francisco poet August Kleinzahler. I'll learn more about it years later, including the fact that a generation elapsed between the poem's genesis and its completion. As he will eventually explain to me over coffee, Kleinzahler began writing it in 1981 and finished it in 2009. For now, I have the poem's music in my ears as I watch the tech bus pass:

> Every evening smoke blows in from the sea,
> sea smoke, ghost vapor
> of lost frigates, sunken destroyers.
> It hangs over the eucalyptus grove,
> cancels the hills,
> curls around the garbage sacks outside the lesbian bar.
>
> And every evening the black bus arrives,
> the black *Information* bus from down the Peninsula,
> unloading the workers at the foot of the block.
> They wander off, this way and that, into the fog.
> Young, impassive, islanded within their tunes:
> *Death Cab for Cutie, Arcade Fire. . .*

Many tech buses are white, like the one I currently have in view. But I know exactly the black buses to which Kleinzahler refers.

Kleinzahler's poem is notable for its interruptions. It begins with the way the fog breaks up the landscape as it rolls over the city each day. Those interruptions are themselves swiftly interrupted by something less natural but just as daily, namely the transport of workers whose presence has, in turn and more permanently, interrupted a prior state of affairs in this city, represented in the first stanza by the lesbian bar. Such bars were a feature of the Mission's 1980s gentrification and were no longer thought of as a feature of gentrification by the 2000s, when they and their regular customers were old guard. As of this writing the last one, the Lexington Club or "The Lex," has shut its doors. The eucalyptus grove Kleinzahler mentions is also a kind of interruption, namely the growth of an invasive species. The eucalyptus trees were originally brought here from Australia. Now ubiquitous and to my eye graceful, they also bring trouble. Very flammable, they contributed to the devastating Oakland Hills fire that began on Saturday, October 19, 1991, and burned fiercely, destroying many homes, before it was contained on the evening of October 20. Eucalypti also release a toxin that can inhibit the growth of non-eucalypti nearby them.

The Mission, north of us, is currently a contested zone, seeming in some ways already lost to its less affluent residents even though many of them are still there, speaking Spanish. I'll be heading all the way across the city for my meeting. The offices of Breakout Labs are grouped with the other Thiel Foundation offices, in a building in the Presidio. This park, located at the northern edge of San Francisco's peninsula, was once Ohlone land, inhabited by that Native American tribe prior to the Spanish settlement, in 1769, of what is now California. Much more recently, the Presidio housed a U.S. Army base (the military planted the trees that now make up the Presidio's forests), and it is now part of the National Park Service. Its protected status as a park has allowed it to remain relatively unchanged, compared to the churn of urban development that is contemporary San Francisco, a churn whose emblematic architectural form might well be the mixed-use condo-minium: retail on the ground floor, with expensive housing units rising two to four (sometimes more) stories above. Climb a hill and gaze across San Francisco's skyline and you see construction cranes everywhere.

There is the churn of change and then there are the churned. San Francisco's ongoing development has victims, and there are rumors of arson in the Mission. Several serious fires have burned down rent-controlled apartment buildings. There is widespread speculation that these fires were set in order to drive out tenants and make way for new condo developments.[11] Many of the long-term residents I chat with seem a little scared, and most of my own friends left the Mission years ago.[12] Uncertainty is its own kind of fog, and wondering if you'll be able to stay in your neighborhood is an everyday form of futurism for many Bay Area residents, especially those who do not work in the lucrative tech sector. Walking in the Mission one day, I see lettering spray painted on the sidewalk, in a familiar Walt Disney font: "The New Mission: Haute yet Edgy!" "This city is just getting better and better," insists one entre-preneur I chat with, who wants to build a fast-food chain centered on leg-umes. While Silicon Valley's dot-com boom of the late 1990s transformed San Francisco, the scale of the current tech boom is far greater, and so is the mag-nitude of the changes it brings. "This city feels like a powder keg," says a friend of mine who works in business consulting and lives in the Mission.

We head through South of Market (SoMa), a neighborhood where tech companies and homeless encampments are equally in evidence. The latter are arguably more evocative of the neighborhood's history than the former. In 1909, Jack London wrote that Market Street divided the tonier parts of San Francisco from "the factories, slums, laundries, machine-shops, boiler works, and the abodes of the working class." There is some historical irony in the fact that SoMa, once a place to hire casual day laborers, many of them homeless, is now the site both of prominent homeless populations, including several tent villages, and of the highly temporary, if often very well-paid, workforce that flows through the tech economy.[13] As of 2013, only the Mission has more homeless encampments.[14] It is important to note that homelessness in San Francisco has only somewhat increased between about 2005 and the day of my bus trip. This suggests that those dispossessed by rising housing costs tend to head to other cities and towns, rather than swelling the ranks of the homeless, although there must be exceptions.[15] The visibility of homelessness in many parts of San Francisco may, instead, have been created by the gentrification of neighborhoods like SoMa, whose new, wealthier residents resent the presence of a homeless population that, en masse if not necessarily as individuals, preexists them.

"What happened to the future?" runs one slogan of the Founders Fund, the venture capital firm Thiel cofounded in 2005, and a separate entity from Breakout Labs. This image of a vanished future reminds me of an observation made by the computer scientist Danny Hillis, in the late twentieth century. Hillis said that the future had been shrinking one year for every year of his life.[16] Every day, he said, he moved forward, but the horizon line of the future, which was made of specific expectations, did not budge. If that line failed to move, this meant that as each specific long-held dream failed to come true, new dreams failed to materialize to pick up the slack. The future was stuck in the year 2000, as if some ineluctable machine, made of progress and generating new expectations, had broken down and ceased to run. "What happened to the future?" implies this same distinctly generational sort of disappointment.

The Founders Fund argues that the world of venture capital has recently been dominated by "cynical, incrementalist" investments, efforts to grasp after quick returns. This, they say, represents a decline from the pattern of "backing transformational technologies" that dominated the field during the 1960s. The function of the Founders Fund statement as public relations copy is obvious, but the almost countercultural provocation—that mere wealth-creation is not enough—still stands. But what purpose could capital have besides growth? Thiel himself has described the problem of how the future seems to be vanishing, saying, "We were promised flying cars but we got 140 characters," a reference to the social media platform Twitter's limit of 140 characters per post. The dream of the flying car began early in the history of aviation. More than twenty years before Thiel's birth (in 1967), a 1946 print advertisement for Alcoa aluminum promised a future in which the personal airplane would be, for the baby-boom generation, what the horseless carriage and horse had been to the last two generations: a means of personal transport. The punning title of the ad, "Your surrey with a hinge on top," references a song from the 1943 musical *Oklahoma!* (in the original lyric, it's a "fringe" on top). Thiel's quotation expresses a frustration with the pace of technological progress, and with its failures to make good on implied promises that once swirled in the popular culture. Cultured meat was a bit of a flying car, too, at the time of Post's burger demonstration, easy to dismiss as a past promise that had never cashed out and probably never would.

Shareable Content crosses Market Street, the diagonal divide of central San Francisco. There's a jolt as we go over the trolley tracks that run down its center. We turn through a couple of intersections, and start the climb up Polk that takes us from central San Francisco up to the Presidio via the historically high-class neighborhood of Nob Hill. My fellow bus-riders are a motley crew. There are a few white-collar professionals who may start their work days a little later in the morning than most, a few schoolchildren, a few plausible retirees running errands by public transportation.

I disembark at the edge of the Presidio. Stately homes abut the greenery of the park, and I can see a sliver of the bay through the trees. The air tastes very fresh and clean. When I arrive at the Thiel Foundation's address, a quick five minutes' walk in, one of the things I notice is that I'm also on the grounds of the San Francisco offices of LucasFilm. The foundation's offices are housed in the same buildings as the film production company. I stroll past a fountain whose four spouts push water past the feet of the wise teacher Yoda, from George Lucas's *Star Wars* universe. In the waiting room, where I check in with a receptionist, I see smaller *Star Wars*–related statues and other mementos of LucasFilm's work. Eventually a staffer leads me from the LucasFilm portion of the building to another waiting area within the Thiel Foundation offices themselves. While the LucasFilm offices are decorated in an unobtrusively corporate fashion, save for the reminders of Lucas's filmmaking career, Thiel's offices are noticeably more luxurious. There's something sleek about the walls. The staffers I see moving between offices are well groomed, neatly dressed, and seem very young. Our proximity to LucasFilm, though, reminds me of the way science fiction films can shape a generation's expectations for technology, the way images of specific technologies (spaceships and "lightsabers," or laser swords, to choose the two most obvious images from *Star Wars*) have come to stand for the future, becoming more iconic than, say, the social realities depicted in those films.[17] *Star Wars* presents us with a seemingly sentient race of robots ("droids") who serve as slaves, and whose labor supports an interstellar democratic republic in its last days and then a totalitarian empire that rises from democracy's ashes, but we remember the droids as technology, and as charming characters, rather than as slaves. In a recent film from the *Star Wars* universe, a character mixes a powder into water, and what looks like freshly baked bread springs into existence immediately. I recall one of Sergey Brin's comments from the cultured beef film: "If what you're doing is not seen by some people as science fiction, it's probably not transformative enough." I cannot help but wonder if, for some venture capitalists, resemblance to science fiction is a criterion for investment.

Hemai Parthasarathy, scientific director for the Thiel Foundation and Breakout Labs, steps out of her office to meet me. She ushers me in to an unused conference room, lets me set up my voice recorder, and we begin to chat. Parthasarathy is generous with her time and full of good humor, which is a welcome counterpoint to my unfamiliar surroundings. Having earned her doctorate in brain and cognitive science at the Massachusetts Institute of Technology, and having subsequently worked in laboratory science, in scientific publishing, and as a consultant, Parthasarathy is an acute observer of scientific research within the academy, of science in enterprise, and of the process by which discoveries become technologies and then companies. I should understand, she stresses, that she doesn't speak for Peter Thiel, but she can certainly speak as scientific director for the foundation. Breakout Labs is unusual, she tells me, among philanthropic organizations. Rather than targeting a problem they wish to solve (examples might include a type of cancer, illiteracy, or childhood poverty), Breakout Labs gives small grants to start-up companies that wish to turn a scientific process or discovery into a technology with (quoting their website, not Parthasarathy) "world-changing" implications. The idea of giving philanthropic dollars to companies whose express intent is to make money may seem strange, but, as Parthasarathy explains, promoting technology development is one of the Thiel Foundation's missions.

Breakout Labs' grants are relatively small by the standards of these things, $350,000 at the most, and, as Parthasarathy tells me, the connections and intangible aid that come with the grant may be more valuable in the long run; notably, many start-up founders will say similar things about their investors. Money is crucial, but so are coaching and networking. While Breakout Labs is not a start-up "incubator" or "accelerator," it does help its companies identify and achieve concrete goals. In exchange for aid both concrete and intangible, Breakout Labs gets a small stake in the company, any proceeds from which then go back into the Thiel Foundation to help with its operating costs. Breakout Labs is thus not a means by which Peter Thiel invests in promising companies. His venture capital entity, the aforementioned Founders Fund, does have the legal right to invest in companies that have received assistance from Breakout Labs, but it enjoys no preferential access to them. The companies keep all the intellectual property they create.

Parthasarathy tells me that a bad habit has developed in Silicon Valley. Investors, she says, often ignore the tractability of the scientific or technical challenges a start-up promises to surmount. Investors are often impressed by the credentials of a company's core founding team, and less than careful when it comes to the science in question. The digital world has shaped this way of thinking, Parthasarathy says. Many investors built their fortunes through software companies, and their expectations for companies are based on the temporality of programming within the controlled environment of the computer. If you put a group of smart young software engineers (well, they don't have to be young, Parthasarathy amends) in an enclosed space with enough pizza for enough time, they tend to solve the software problems you give them.

On the other hand, Parthasarathy points out, presumptions about the tractability of scientific challenges run through nonprofit and academic environments too. The amount of money available for autism research, for example, has shifted many researchers into that area, although, as Parthasarathy says, "the science may not be there yet" to produce therapies. We may not have the necessary knowledge of how the brain works, so all the research funding in the world may not yield a fast answer, and it certainly won't produce timely therapies for those who badly need them. This does not, she notes, mean that autism research is bad science, but merely that it can be frustratingly incremental. Parthasarathy observes that within the start-up world, biologists are sometimes seen as having a "defeatist" attitude or as lacking "the attitude of success." Soma, I reflect—meaning the body and its cells—presents harder problems to solve than code (and so does SoMA, and San Francisco's gentrification crisis). This supplies one answer as to why Breakout Labs doesn't direct funding according to specific areas of medical or social need. It allows them to be flexible and to acknowledge when problems appear soluble and when they don't.

Modern Meadow, Parthasarathy says, was "pretty damn close" to being a perfect grantee for Breakout Labs. As she puts it, they are trying to solve nontrivial problems in tissue engineering, and they also have a larger vision of changing how the world consumes animal proteins. The father-son team of Gabor and Andras Forgacs has already founded another company (Organovo, which seeks to create organic tissues for drug testing), thereby demonstrating their entrepreneurial skills. They also have the commercially strategic idea of creating leather first, rather than beginning with the lofty goal of meat. Not only is leather—potentially just a thin sheet of cells—easier to grow than meat, it also sells for a much better margin, particularly because Modern Meadow plans to begin by working with the high end of the fashion industry. Other grantees in the first years of Breakout Labs' operation have included Skyphrase, a company that builds tools to help scientists sort data based on natural language, and Stealth Biosciences, which builds nanoscale devices for biological applications at the cellular scale, in both research and therapeutics.[18] While all these companies may have different goals, including (needless to say) making money, it is easy to see that all their projects advance the Thiel Foundation's larger mission. "Technology," Parthasarathy says, "is seen by the foundation as a way of empowering personal freedom. So technology is viewed by us as an overwhelmingly good thing, as a way civilization advances." But, she stresses, to be oriented toward progress is not necessarily to approach it "top-down," with particular end states in mind. I make a mental note regarding the tension between these approaches: to approach the future by encouraging the creation of technologies, while trying not to become too attached to particular visions of the future.

I thank Parthasarathy, gather my voice recorder, and say goodbye. When I leave the building it is midday, and the fog appears to be on hiatus. I hope it returns soon. I muse over the one issue we did not discuss directly, namely cultured meat's relationship with regenerative medicine. Cultured meat essentially borrows tissue-engineering techniques from the world of regenerative medicine, in which scientists culture cells in hopes of ultimately creating functioning organs and other tissues for grafts into live patients. Many of the companies funded by Breakout Labs target challenges in medicine and health; some of them do pursue an aim in which Peter Thiel has expressed direct interest, namely the extension of the healthy human life span, an aim that has garnered considerable (and usually very critical) media attention at the time of my visit to the Thiel Foundation. There is a common term for the effort to use technology to transform the human condition: "transhumanism," originally coined by the evolutionary biologist Julian Huxley (brother of Aldous, author of *Brave New World*) in 1927.[19] He understood it as the use of science and technology to explore the possibilities of the human condition, which included transcending its native limits. As I walk out onto one of the many hiking trails that wind through the Presidio, I think of Post's claim that a tiny biopsy of cells could produce tons of meat, and I wonder if cultured meat is transhumanism for beef, a desire to go beyond the physical limitation of particular edible species.

Kleinzahler's "Hollyhocks in the Fog" ends on an ambiguous note, as the poet refuses to pass final judgment on the tech bus or the workers it carries. "There is nothing further to be known," he writes, referring to the impressive power of Google, the Internet search engine, to return answers to a plethora of queries ("Ryne Duren + wild pitches + 1958"; "Huitzilopochtli"). But these answers have no effect on the flow of the day:

> The fog, like that animate *nothingness*
> of Lao-Tzu's sacred Tao,
> has taken over the world, and with night settling in,
> all that had been, has ever been, is gone,
> gone but for the sound of the wind.

Has the fog taken the world back from the tech workers? This would be a reading of the poem very congenial to many people struggling to survive in the current version of the Bay Area. But in fact the fog has displaced absolutely everything. It's a leveling force that can mock human aspirations that include both "progress" and "justice."

Later in the day, I sit on my friend's deck in Potrero Hill, enjoying a privileged view north toward downtown San Francisco as the actual fog starts to reclaim the city, as I had hoped it would. Soon everything will be made of fog again. I get less a sunset than a sun-swallow as the fog takes the world back. The Bay Bridge disappears, and the white-gray swallows an old wooden church perched on the hill. There must be tech buses bringing workers home to Potrero Hill, but I can't see them. My visit has told me only a little about the politics surrounding cultured meat, and a little about where funding for it comes from. But there has been much to learn about the contextual fog out of which it emerges.

While funding from the Thiel Foundation says little about the politics of cultured meat, the relationship between cultured meat and the world of small start-up companies will be crucial to the story of this technology's emergence, as will the assumption, eyebrow-raising for some social scientists but widespread among cultured meat workers, that market forces are a reliable means for producing positive social change. Assuredly, market forces do change society all the time; this is capitalism in its guise as a historical actor.

But questions about food are not only questions about markets, but also, for many, matters of social justice, public health, community dignity, and much more. They are intrinsically political questions, not only about how our food is produced, but also about by whom and for whom, and about our choices in dining companions. And they are questions about the role of the state in the sustenance of its people—for, and perhaps contrary to the wishes of some, in modernity the state is the central and paradigmatic political form.

To claim that the present state of technology is disappointing compared to past visions of the future, as Peter Thiel has claimed, is partly to observe that our physical technologies (in energy, in transport, in medicine, in manufacturing) have lagged behind our digital ones. It is also to suggest that the digital has disappointed us compared to earlier visions of the future of computing. Cyberspace has proven less liberating and productive a form of innovation than was once predicted. All of this prompts a set of obvious questions, the trunk and legs of an elephant in the room of the fog: Who got to set the terms of progress anyway? What money bought them that right? And what were their politics? Somewhere in the mix floats the problem of living tissue, so much harder than code to understand, to build, to (I am learning to use this as a verb) "scale." The future has been shrinking every year of my life, too.

FIVE

Doubt

Over five years I gathered a lot of doubts. They ran from heavy criticisms of the feasibility of cultured meat, made by scientists, all the way to laypeople's gripes left in the comments sections of online news articles about cultured meat. To be sure, doubt, like hope, reflects bias. A scientist with a company that produces plant-based hamburgers insisted that cultured meat was a dumb idea, impossible to scale up.[1] An anonymous commenter said cultured meat was disgusting and "unnatural." As cultured meat attracted more and more media attention, a celebrity chef appeared in a very short Internet video in which he complained about "fake" meat and proclaimed his fidelity to the "real" thing.[2] Doubt appeared at cultured meat conferences, in exchanges between science journalists and researchers—albeit not among entrepreneurs, for whom a confident showing was crucial. As I spoke at conferences myself, journalists reached out to me (very few "experts" were available), and their questions invited me to doubt: Did I think cultured meat was viable? If so, when would it arrive? I insisted that these weren't the questions I was trying to answer, and I frustrated a lot of journalists. My own questions, I said, were the sideways ones of the anthropologist or the historian, unlikely to provide much hope. When venture capitalists called me, I insisted I was an implausible source of business advice. Meanwhile, doubt could be found in survey data: a 2014 Pew Research Center report on American attitudes toward the future of technology found that only 20 percent of respondents would eat meat "grown in a lab," as the survey put it.[3]

Despite my evasive public answers, I felt both hope and doubt. They cycled like the moon as I did my research. I'd get a positive report from a lab; I'd get a negative report from a lab. It is hard to gauge ultimate technological success from a single promising experimental result. Millions of dollars of funding would flow in from a famous American billionaire's foundation, or a prominent Hong Kong venture capital firm, and I would wonder if there was any meaningful sense in which investment guaranteed progress. I strongly suspected that there was not. Sometimes technical roadblocks don't go away. As Hemai Parthasarathy told me when I visited Breakout Labs, the problem with targeting specific scientific or medical problems and flooding the appropriate labs with cash is that not all problems can be fixed with the things money buys. And yet it was impossible not to feel a sense of elation during weeks when the field showed signs of real progress.

From the start of my research in 2013 to its conclusion, two technical roadblocks increased my doubts about the viability of cultured meat. These were finding a serum-free growth medium and producing three-dimensional or "thick" tissue. These were not the only potential roadblocks—others include controlling the temperature within bioreactors and building appropriate scaffoldings or microcarriers onto which cells can attach—but they are the ones most often discussed, perhaps because they seem the most difficult to surmount. While many commercially available growth media are free of the decidedly non-vegan fetal bovine serum, they have all been judged to be too expensive for use at industrial scale. As regards dimensionality, growing cells in sheets is far easier than growing them in three-dimensional forms, but those latter forms are needed to reproduce the layered muscle of more complex meats, such as steak. When I thought about the complexity of painstakingly replicating something already available at a restaurant on the corner, I felt fresh doubt.

Tissue engineers speak of the nutrient and oxygen diffusion limit, or the distance from blood supply a mammalian cell can live. It is about 100–200 micrometers, about the width of the widest human hairs or the thickness of most paper. This means that bioreactors for the growth of three-dimensional tissues require vascularization, or artificial blood vessels and veins. Tissue engineers working in regenerative medicine have put in long hours trying to create good vasculature systems, in order to grow tissue for transplantation, achieving limited success. The medical implications of a perfected vasculature system for mammalian tissue would dwarf its implications for growing meat. As a biomedical engineer said to me at a cocktail party, a tissue-engineered heart has been ten years away for thirty years.

In 2012 the synthetic biologist Christina Agapakis, then a postdoctoral fellow at the University of California, Los Angeles, published an editorial in which she described scale as the bugbear of cultured meat (which, in keeping with 2012 conventions, Agapakis called in vitro meat).[4] She referred to scaling as the "*deus ex machina* of many scientific proposals," the problem that scientists want to pass on to the engineers who implement their ideas. Scaling seemed a great deal to ask of tissue culture, given the sheer expense and technical challenge of creating even small pieces of mammalian tissue. The costs of growth media, heating, technician time, and so forth seemed staggering. The magic wand of scale, Agapakis added, had been waved over other foods in the past. In the case of algae, touted as a solution to global malnutrition and "Malthusian catastrophe" in the 1950s, the transition from laboratory conditions to industrial conditions proved fatal.[5] Certainly, when considering the potential future of technology, the past is an uncertain guide. There are plenty of past technological failures to choose from, and a precedent does not segue gracefully into a forecast. Yet *Chlorella pyrenoidosa,* the species of algae in question, seems an apposite citation. The chief problem was that *Chlorella's* remarkable ability to turn sunlight into food led to eager extrapolations from small scale to large, without attending to the details of the latter. An article in the *New York Times* promised protein-rich yields about a hundred times as great as conventional crops, per acre, because a feature of the organism's biology had been taken as a guarantor of plenty. This assumption obscured the difficulties that would be encountered in making *Chlorella* a food source (centrifuges were involved, due to *Chlorella's* small size). The economics of large-scale *Chlorella* production proved impractical even as other crops outperformed expectations for agricultural yields, allowing their prices to remain low. Once, when I stepped into the arrival hall of Kansai Airport outside Osaka, Japan, I saw a poster that advertised *Chlorella* as a dietary supplement. A pleasingly light green powder spilled out from a white vial.

In a second editorial published on August 6, 2013, Agapakis's skepticism was undiminished by Mark Post's widely publicized hamburger demonstration the previous day.[6] She referred her readers to her earlier doubts and raised new and more manifestly ideological ones, citing Ursula Franklin, a physicist and metallurgist best known for her writings on technology. The salient distinction in Franklin's work is between "holistic" and "industrial" models for the growth of things both organic and artifactual. "Growth," Franklin says, "cannot be commandeered; it can only be nurtured and encouraged by providing a suitable environment."[7] Production models are "different in kind" from natural models. One of the chief defects of the former is that they tend to ignore externalities, or the effects of production on the world beyond "a work situation, a production line." Franklin writes,

"We know that the deterioration of the world's environment arose precisely from such inadequate modeling." The message is simple: techniques for making things cheap by making them scale have tended to shift costs to the natural world, although Franklin, a Quaker, is careful to mention the deterioration of the physical and mental health of workers as well. Franklin's reasoning may not turn out to be final and authoritative for the future of meat, if (against all doubt!) cultured meat does manage to scale successfully, and more sustainably than animal agriculture has done. Yet, as Agapakis's invocation of Franklin shows, doubt can be productive. Here it yields a vision of organic things sitting uncomfortably within industrial modernity. That very discomfort may explain why so many cooks and chefs turn away from processed foods and toward the organic world with an almost animist enthusiasm, trusting the grown more than the made. Franklin might want us to ask in what precise sense the line between the grown and the made could be erased, and with what consequences. After all, cultured meat's pioneers would like to scale quickly. They wish to move, in very little time, from an artisanal mode of producing small amounts of muscle tissue by hand to a veritable industrial revolution. Franklin's "growth" appears to be the more gentle and gradual process that follows birth, typically a process sheltered and shaped—Franklin writes "nurtured"—within those comparatively slow-moving things, human community and culture.

SIX

Hope

Are open hands clean hands? I followed the story of cultured meat because it spoke to a complicated corner of my soul. I had long been troubled over carnivory. Is it ethical to eat animals? If eating animals is unethical, but I continue to eat them (reader, I eat them), what does it mean to live with that hypocrisy? These are not just questions about animals and their moral status, but about our appetites and our prospects for moral improvement, both as individuals and as a species. I have caught myself feeling distinctly uneasy when handling raw salmon or pork. To restate one premise of much discourse around cultured meat, our appetites, whether they come out of our cultures or out of our animal natures, may be inimical to our desires for a better world. If hope is a posture toward an outcome judged to be neither fully certain nor beyond all possibility, then our own appetites stand as the literal prospect from which we survey potential global futures. I had always wished to live in a better world—a basic impulse that leads to utopianism, philosophizing, disappointment, and to reading science fiction, the latter being a very common interest in cultured meat circles. The idea that technological progress may aid moral improvement is also widespread.

Notably, the phrase "moral improvement" places a moral actor in the spotlight rather than focusing on the effects of her deeds. My use of the phrase tags me as something other than a consequentialist, in the terminology of moral philosophers; it hints at my concern with the character or virtues of actors, whereas many supporters of cultured meat are more concerned with actions and their outcomes. Following in the footsteps of many utilitarian philosophers by considering animal suffering worthy of moral regard, they hope cultured meat will reduce the total suffering experienced on this earth.

Reasons for hope are often harder to substantiate than reasons for doubt. When it comes to scientific or technical projects, doubts assemble in the form of discrete bits of evidence that pile up, whereas cultured meat's supporters hope for something whose plausibility is inherently hard to establish. They reach not just toward technical success but also for an outcome that would necessarily be mediated by markets and consumer choice, and that would mean replacing one massive form of infrastructure with another. When I conducted interviews with cultured meat researchers or with outspoken supporters of such research and asked about their motivations, the most impassioned personal stories I heard centered on animal protection, with environmental defense trailing as an intellectually serious but less fiery drive. Although food security and the defense of human health are also reasons to move away from industrial animal agriculture, my interviewees rarely spoke of such things.

One young cultured meat researcher who had worked as an assistant to a large-animal veterinarian painted me a grisly scene. A farmer's cow had an infection in one eye that threatened to spread. To save the animal, which was valuable property, the eye had to be removed. The catch was that the animal was worth less than the cost of anesthesia. And so the young researcher, who was studying to be a veterinarian herself, was tasked with holding the cow as the eye was cut out without anesthetic. The operation took hours. The cow, valuable enough to live but not valuable enough to be spared suffering, made her feelings obvious. Know the hope for cultured meat as the hope for an end to such things.

The early scholarly literature on cultured meat reads like a field guide to hope. During my years in the field, scholars of science, technology, and society, anthropologists, bioethicists, and others published articles in which they affixed labels to cultured meat and to its creators and celebrants. Their efforts were mostly descriptive, as befits a novel topic of study, but with the effect that much of this literature reads like articulate restatements of the cultured meat workers' own sentiments. The cultured meat movement has for the most part been allowed to set the terms of the conversation about cultured meat, and to determine which questions about cultured meat are the most important ones. In one essay a two-author team catalogued the apparent advantages and dangers for morality that the technology seemed to hold.[1] Upholding the most obvious advantage, an end to widespread animal suffering, the authors suggested that this outweighed the possible downsides, which include the ambiguous "harm" of disrespect to the integrity of animal bodies and to their natural form and the danger of our growing hubris as biotechnology reinforces human domination over nature. These are commonplace criticisms in the world of medical bioethics, anxieties readily transferred to the emerging domain of food production by tissue culture.

Most startlingly, the authors suggested that developing cultured meat might be seen as a moral obligation. This reversed the normal order of operations for bioethicists, who normally respond to existing new technologies and their prospects for human health and ethics. "Morality is not something that must simply respond to new technologies as they arrive, throwing us into confusion," they argued; rather, "morality may champion and assist in the development of new technologies." This, in turn, would be "a step toward the production of a world that in fact, and not merely in ideal, mirrors the moral vision we possess for it." Expanding on this latter point in another exemplary essay, a different two-author team wrote that despite widespread characterizations of technology as a deadening influence on modern life, new technologies like in vitro meat production could "disclose" (their term) worlds of ethical possibility, in which publics might convene to contemplate their collective moral choices.[2]

These two essays manage hope in slightly different ways. The former, published in 2008, years before Post's hamburger demonstration, argued that a moral ideal could appear and then effectively demand technological implementation. The latter, published in 2012, also before Post's demonstration but with the cultured meat hype cycle on the upswing, captures a sense of the dynamic interplay between signs of a new technology's emergence and the reciprocal transformation of moral debate and decision making around that technology. What seems clear is that such reflection on technology and ethics flatters the novel object of study, and cultured meat has been lucky to find commentators who find its proposed coordination of morality and technology appealing.

Neil Stephens, a sociologist of science and technology who has been tracking cultured meat since the mid-2000s, captured the preliminary nature of all these writings when he called cultured meat an "as yet undefined ontological object." The sheer reality of cultured meat, Stephens reminded his readers, still remained in question.[3] Meanwhile, other social scientists began to ask a less philosophical but equally significant question by means of surveys: Would European and North American consumers reject cultured meat out of hand, as an "unidentified disgusting object," or would they be willing to try it? I could not help but wonder about the effects of such surveys on the people taking them. Even if they presented cultured meat as a hypothetical thing, not yet realized, how could researchers avoid giving the impression that they were, effectively, heralds of a very real new food?

"For what may I hope?" asked Immanuel Kant, who was, not at all incidentally, a moral philosopher of duties and obligations rather than a consequentialist. How can we hope without imagining there to be a discrete and possible outcome for our hoping? Kant's interpreters have found many meanings in his account of hope, including seeing hope as a species of religious faith, or as something to which we may have a moral obligation, but what seems certain is that Kant understood hope as a way of relating to the uncertain prospect of a non-impossible desired outcome, such as moral improvement. Philosophers vary in terms of how rational they understand hope to be. Kant found ways to adequately reconcile hope and rationality. For Søren Kierkegaard, writing generations later, what mattered was for hope to leap beyond reason. And the twentieth-century Marxist philosopher Ernst Bloch created an entire philosophy of hope that sought to attune us to all that is metaphysically possible in the future, in contrast to philosophy's traditionally retrospective role. However, a common sentiment is that hope's role is to help us move beyond experience and into possibility. The question I am still turning over has to do with something more discrete, namely the relationship between a radical reduction in the suffering of animals and the improvement of our moral character. Does hoping that cultured meat might lead to the former mean that hope for the latter has been worn through and abandoned? Have we forgotten the old dream of improving our moral characters and replaced it with the dream of a new kind of prosthetic technological morality?

Hype, not doubt, may be the real enemy of hope in the case of cultured meat, but (ironically enough) hype may also be necessary if cultured meat is ever to materialize fully. A number of scholars who examine emerging technologies converge on this point: press releases are performances, often made in search of funding, and thus hype is "constitutive" for bringing the future back to the present.[4] In other words, it is promissory. Of course, the necessity of hype does not mean that hype is safe. Hype can inspire hope, but if in the fullness of time hype seems unjustified, hope will be dashed, perhaps permanently, at least within a given group of supporters (a "community of promise") and for the company or individual that has been encouraging hype. There are business consultants who call this effect a "trough of disillusionment," with the expectation of eventual recovery and a return to productivity, but, as many of the people I interviewed suggested to me, a widespread sense of broken promises could easily end cultured meat altogether. Meanwhile, hype likes to think it is something less than a firm promise and tries to slide beneath the bar of accountability, the bar against which hopes and promises are eventually judged. If hope is necessary it is also, as the literary critic Fredric Jameson puts it, "the principle of the cruelest confidence games and of hucksterism as a fine art."[5]

SEVEN

Tree

I just finished my visit to Breakout Labs, at San Francisco's northern edge. I'm tempted to spend the rest of the day wandering around the city and thinking over Breakout Labs' challenges: how to determine what kinds of problems in science and engineering are tractable, how to know what bets to make in pursuit of progress, how to define progress itself. But a very tall wooden structure, across a clearing, grabs my attention. And it wants more than the time of a middling-sized mortal mammal; it seems to angle for the sky. Closer inspection introduces me to *Spire*, a 2008 creation of the English artist Andy Goldsworthy, who works in natural materials. Based in Scotland, his atelier's reach is international, as is his reputation. His works currently dot the Presidio, sometimes blending into the park, sometimes erupting from it. *Spire* is a bundle of tree trunks that stands nearly a hundred feet tall at its tip, an artifact made of organic matter. A few trail runners streak past.

One of Goldsworthy's central thematic interests is time. In the 1980s, he sculpted elaborate arches from ice. He once had to piss on a stone support in order to free it from an ice bridge that had grown solid enough to stand on its own.[1] Goldsworthy's decision to work in the medium of frozen water tipped the theme of time toward the subtheme of impermanence, but in later work he reminded his audiences of impermanence through one of our longer-lasting building materials, stone.[2] His 2005 *Drawn Stone* is a crack in the paving stones that lead up to the door of the de Young Museum in San Francisco's Golden Gate Park. This crack recalls the fault lines that run through California, as well as San Francisco's constant and permanent vulnerability to quakes. A slab of stone stands in the crack's path, seemingly cloven in two, and serves as a double bench. This is humor dark and playful, inviting us to rest in danger.

Goldsworthy made a much smaller precursor to *Spire*, called *Sticks Spire*, in Cumbria in 1983, an almost human-scale project that one sculptor could complete on his own. Decades later, *Spire* took a large team and heavy machinery to construct, though it juts from the earth as if it grew there. The *Spire* team built the sculpture by bundling together the trunks of thirty-seven Monterey cypress trees, which had been cut down as part of a reforestation effort in the Presidio. A hundred feet tall at its peak, *Spire* will eventually be obscured by the ring of Monterey cypresses that now grow around its base. It is hard to believe this, looking at the ring of trees now. They are still juveniles, their branches moved by the slightest wind. Some experts believe these trees can survive for up to two thousand years. Others dispute this, saying that the oldest specimen yet found was just a few hundred years old.

Spire is linked to the reforestation efforts from which its wood comes, and reforestation will eventually hide it, assuming that civilization endures long enough. *Spire* speaks to environmental worry, which thinks in long timescales in relation to the pace of technological development and investment. Financial quarters are nothing to a tree, though this is easy to forget in San Francisco—quite an irony, for the city has been a historic stronghold of the environmental movement far longer than it has been a tech hub, though it's been a gold-rush town for longer still.[3] As I stand here in the Presidio, questions about emerging technologies, companies, and the future-oriented gamble of investment all stand juxtaposed against the complexities of Goldsworthy's work: time and growth, time and impermanence, and the risks nature holds out for civilization—and perhaps for Californian civilization in particular, for reasons tectonic.

EIGHT

————

Future

A grasshopper covered in chocolate adds an entry to my life list of atypical eats. While I've eaten them cooked Oaxacan style (as *chapulines*) in Los Angeles, I've never eaten an animal transformed into candy. Nor have I eaten an animal nominated by its promoters to resolve food security crises worldwide.[1] I'm standing around a small table with my fellow participants in a workshop on the future of food. We're eating insects, not just grasshoppers but also ants and mealworms, during what amounts to a product demonstration by insect entrepreneurs, their neat stacks of business cards on the table. Grasshoppers are very expensive to harvest compared to the costs of industrial-scale beef, but some think they could be raised at scale rather than gathered from fields. In theory they might be far less resource intensive to produce than cows, pigs, or chickens.

It is November 2013. We've spent the past two days in a medium-sized event hall in downtown Palo Alto, listening to stories about the future of food and getting the occasional object lesson like this one. We—some thirty of us, balanced between genders, mostly white, mostly middle-aged, many of us employed in the R & D or strategy offices of major food companies, all of us dressed in slightly coffee-stained "business casual"—have the buzz and fatigue of people who've spent a lot of time talking excitedly in an enclosed space, a new topic every fifteen to thirty minutes. We've heard about gut microbiomes, public nutrition education, and edible sensors that send out data even as they're digested. I'm curious about what motivates a busy managerial type in the food industry to turn up at a workshop like this one, broadly educative but not designed to produce specific "deliverables" for clients. The grasshoppers are delicious; the energy bars made from mealworm flour are not. If we really are facing a near future in which insects are a common source of animal protein in the developed world, then I hope it's a future of whole bugs, not bars. It's an intriguing possibility, because there are very few animals that North Americans like myself eat whole. Most of us can manage a grasshopper in a single bite.

This experiment in entomophagy is part of "Seeds of Disruption: How Technology Is Remaking the Future of Food," a weekend workshop organized by the Institute for the Future (IFTF) in Palo Alto. IFTF is a combination consulting firm and think tank that has been doing "futures work," as practitioners say, since 1968, when IFTF was founded near Wesleyan University in Middletown, Connecticut, by Rand Corporation researchers working with a grant from the Ford Foundation. It moved to Palo Alto soon thereafter, securing the organization's future ties to the technology sector and enabling staff to ride their bicycles to work if they so choose, the whole year round. IFTF was founded at the leading edge of a decade of growth for think tanks, many of them examining the future of food, many of them informed by demography, environmental science, and troubling statistics regarding global population growth and projections for the future of sustenance.[2] IFTF was not centrally concerned with food at its founding, though today food is a prominent part of IFTF's portfolio.

The futures of telephony, housing, and newsprint—technologies with immediate and critical effects on society—were concerns for IFTF in its early years, just as human-computer interaction and virtual reality are concerns as of this writing.[3] IFTF clients and partner organizations range from the Rockefeller Foundation to the U.S. Department of Naval Research to the food giant Hershey's. One of its largest events is the annual "Ten-Year Forecast" conference, which helps a large number of delegates think about global issues as they might stand a decade out. One year the event grew so large it was held on an aircraft carrier harbored on the east side of the San Francisco Bay. I'm here at this much smaller event to glimpse work on the future of food, and to try to understand the marketplace in which ideas about the future of food are bought and sold. Cultured meat is just one such idea; bugs are another; urban agriculture in the heart of cities is yet another. The three are often crammed together at conferences under that capacious banner, "The Future of Food."

While some futures workers do sell predictions, including ones about technology, there is little agreement among futurists regarding the possibility of knowing what's to come. "We can't predict the future," says IFTF's executive director, Marina Gorbis. A poster spells out IFTF's response to this problem:

the words "I have seen the future" are printed with a slash through "have seen," and "am making" is written in their place, although I will learn that this gesture captures only one of IFTF's organizational moods. Besides prediction, other common tools for futurists include forecasting and working with scenarios. We can distinguish between these three tools as follows: if prediction describes a specific set of events that will take place (it will snow tomorrow), forecasting gives that set a probability (there is a 30 percent chance that it will snow tomorrow). By contrast, scenario work describes a set of conditions that might come about, are tough to quantify, and have certain consequences (if it snows tomorrow, will you snowshoe to work?). Professional futures workers may specialize in one mode or move between them, and depending on their orientation they may object to drawing hard lines between modes.[4] Some are frank about how their approach is shaped by their clients' demands. Futures work is, by and large, more a form of consulting practice than a scholarly discipline, and it is to be distinguished from the future-oriented pursuit of specific utopian visions. While some futurists find work with city, regional, or national governments, or with non-governmental organizations, the primary client pool is in business. There is all the difference in the world between the history of philosophical speculation on the future and futures work, though the latter sometimes borrows dignity from the former.

As I spend time at IFTF, I learn that the staff is far from homogeneous in their educational backgrounds and aspirations. Some are attached to the label of futures work as a source of professional identity. A few even hold advanced degrees in futures studies, from the few programs in that field that exist, such as the Hawaii Research Center for Futures Studies at the University of Hawaii, Manoa, directed by Jim Dator. Others, especially younger staffers just out of college, are less attached to the title of futurist and see IFTF as a training ground for other eventual pursuits, ranging from graduate school in the social sciences to activism or work in non-governmental organizations. During my visits to IFTF, I spend most of my time with the staff members who make up the food team, all in their mid-twenties to early thirties. IFTF runs young, which I gradually learn is common among consulting firms and some think tanks. There is talk of "T-shaped people." This means people with deep vertical specialization who branch out, as it were, at the top. I notice that T-shaped people often wear blue jeans, in contrast to us attendees.

For the food entrepreneurs who brought their bugs, this IFTF workshop offers a chance to provide samples to a group of potentially influential people who work in the food industry, often in large companies with impressive economic and environmental shadows. For IFTF, it is an opportunity to use a local entrepreneur's wares in a storytelling exercise, giving their clients an object lesson in one possible future for our food supply. The grasshoppers are not an IFTF prescription, but a visceral way to get the mind moving. A visiting executive from the research and development office of a major soda company, sitting to my left, remarks that he expected more concrete predictions about issues affecting food supply and food security, such as climate change. I nod, chewing, concurring. I ask him what major concerns of his company brought him here. "We're trying to reach a health-conscious market," he says, making me think of a poster on New York City subway cars that illustrates the number of sugar packets in a typical twelve-ounce soda. "And we're worried about water," he adds. Against any expectations that the Institute for the Future's name might create, they deal less in predictions than in scenario work, informed but not necessarily governed by the sense of probability that the word "forecasting" carries. As of 2013, IFTF's keyword for describing their version of futures work is "foresight." Despite their extensive work with corporate clients, they are categorized as a nonprofit organization on the argument that the cultivation of foresight—in communities, in governments, and in business—qualifies as a social good. What foresight produces, according to IFTF house publications, is not knowledge of the future, but better-prepared actors and, in a civic key, better citizens. "Foresight," the Dutch sociologist, futurist, and social-democratic politician Fred Polak wrote, "presupposes a conception of time, duration, development, and continuation," but it also means preparedness for crisis.[5]

Passersby in downtown Palo Alto often mistake IFTF for a branch of Stanford University. This captures something of the feeling of this town, where the line between academic research (especially in engineering and the life sciences) and industry can be hard to see. I remember one technology journalist, in a moment of carelessness, referring to a "Stanford executive" rather than a "Stanford professor," when the latter was meant. Posters on the wall offer futurist quotations, such as "The future started yesterday, and we're already late" or "Any useful statement about the future should at first seem ridiculous," the latter from Jim Dator. A bookshelf holds books by IFTF staff or affiliates; there are books on technology and cities, books on social networks in the business world, books on leadership, books on video games. "We are called to be architects of the future, not its victims," says the late architect and polymath R. Buckminster Fuller in text printed on one window.

Fuller was a great influence on mid- to late-twentieth-century futurism, but he is perhaps most famous as the designer of the geodesic domes that fascinated many members of the 1960s counterculture. They became a staple of the architecture at some communes, such as Drop City, which persisted in southern Colorado from 1965 through the early 1970s. Fuller hypothesized a type of person who would operate parallel to, but separate from, the development of new technologies and imagine their adaptation as useful tools to fulfill human needs; he called this figure the "comprehensive designer," and there is a certain affinity between such a person and a consultant like the ones working at IFTF—not the originators of ideas in technology or policy, usually, but the people who see the potential of those ideas and serve as the agents of their adoption. The architectural historian Simon Sadler, who notes that Fuller's authorship of the domes is not uncontested, sees the domes as representing something much larger than mere utility: "an authorless mathematical certainty" that "connect[s] the geodesic dome's builder-occupant to patterns underlying nothing less than cosmological order."[6] To be an architect of the future, after Fuller, might mean tapping into forms beyond human contrivance, creating systems that reflect not only our pursuit of pragmatic ends but also our reach for ideals. We might make representations of the universe's fundamental intelligibility, or, more modestly, pitch tents that seem like arguments for that intelligibility. Such cosmic matters might be distracting if one were seated in a chair at IFTF, trying to write a report on the future of water resources when Fuller's quotation catches one's eye, but the gains of an inspiring semiotic environment are worth the cost. On the best days the physical space of IFTF feels like a machine for creating useful associations between discrete conceptual units.

"Foresight" seems an intentionally vague word to use to describe an organization's mission. This may reflect the nature of consulting work itself. While some consultants produce concrete deliverables, such as design schematics or financial instruments, others provide services that are less easily quantified. In practical terms, "foresight" seems to mean the ability to plan, embrace contingency, and think about possibilities in a way that many organizations are not attuned to doing. Although some business consulting practice is founded on the notion that the consultant possesses topical expertise that the client lacks, such as how to enter the Brazilian chocolate market or what kind of whiskey young adults are likely to order in a bar in Taiwan, IFTF works differently. They attend to a wide field of "signals" of change that their client organizations do not have the resources to observe but find it very helpful to know about for strategic planning purposes; signals are basic units of operation in IFTF futurist exercises. By working in ways their clients ordinarily cannot due to their lack of time and attunement, IFTF helps them prepare for uncertainty. It also exploits a certain strain of business school wisdom according to which innovation, by its very nature, cannot occur within established organizations, which are bound to please their existing shareholders and customers.

In a video promoting their work on food, IFTF staff members speak of "catalyzing the innovations that will remake the future of food" and of "food innovation hubs"[7]—places, presumably, where new ideas in food come from. I immediately think of Copenhagen in its current guise as a destination food city, home to some of the most inventive restaurants in the world. IFTF nominates its own town, Palo Alto, situated not only next to Silicon Valley but also close to California's agricultural heart, the Central Valley, which happens to yield up more than 50 percent of the nation's produce. Between November 2013 and 2015, I will visit Palo Alto several times, as a participant observer in IFTF workshops like this one, as a guest speaker, and as an audience member at public events. It will be with IFTF's assistance that I make many of my connections in cultured meat research, becoming part of the trading zone between big industry, entrepreneurship, and academic expertise that IFTF makes possible.

One of the IFTF staffers enjoins us, "When you introduce yourself, please name a technology that gives you hope or fear for the future of food." Thirty-six hours before the *chapulines,* we go around the room and make introductions at the beginning of the first day. We've already written down our hopes or fears on a large white sheet of paper with our names next to them, and a Polaroid picture of each of us has been taken and taped down next to our contribution. Many people mention genetically modified organisms in a hopeful key, and just as many mention them in a fearful one. I name cultured meat, and I say that I feel both hope and fear, not knowing if it represents a positive or negative future for animal protein. IFTF staff members have just welcomed us with an introductory talk about the institute, its methods, and the main themes of the workshop. Miriam Lueck Avery, codirector of the Global Food Outlook program (later renamed the IFTF Food Futures Lab; other "labs" include the Emerging Media Lab and the Governance Futures Lab), now has the task of walking us through one of the lavishly designed handouts that are common at IFTF workshops. This is a large map, twenty-five by twenty-two inches, dominated by a radial design. Five sections are divided into green, yellow, blue, purple, and brown: "manufacturing," "distribution," "production," "eating," and "shopping," all very legible as activities central to our food system. The sections are then subdivided into other sections working out from the middle: "core strategies," or general vectors for change that might be expected in that sector; "disruptions," or, in IFTF's version of Silicon Valley parlance, a technological change that suggests a new direction for a sector; "strains of uncertainty," which make up the last section, are described as low-probability developments that, if they do occur, could change everything. Within the "production" pie wedge, the relevant "disruptions" include "growing food on every surface" (a staple of urban agriculture), "swarming robot farmhands," "reformulating eggs," and "fooling food critics," the latter being a reference to a company that uses plant proteins to create hamburgers nearly identical to meat, effectively a competitor technology to cultured meat. At least one food critic reports being fooled by the company's burgers.

Some of the disruptions or strains of uncertainty listed, I note, are already in effect, including grazing practices meant to restore grasslands. Others, such as drone delivery of groceries, or on-site manufacture of goods via 3D printing, have been prototyped and fed into the same tech-media hype cycle as cultured meat. One IFTF publication dwells on the idea of a new regime of production such that consumer goods might be created only on demand. Call this the "matterstream." It would rely on digital files that can be printed or otherwise fabricated on demand, so that most objects exist in silico until they are printed. The savings in energy and environmental wastes, to say nothing of natural resources, might be enormous, but then again, we are under the spell of the hypothetical. As we grabbed breakfast in IFTF's main hall, before the hopes/fears exercise, David Bowie's song "Space Oddity" was playing on the sound system. "There is a virtual reality headset at the back of the room, for anyone who wants to try it out," we were told. I had forgotten that "Space Oddity" opens with the lyrics "Ground Control to Major Tom /Take your protein pills and put your helmet on."

The "Seeds of Disruption" map has a visual logic. At the map's center are the predictable activities of farming, food production, marketing, shopping, and eating. As the eye moves out to the map's further edge, it encounters increasingly improbable developments. But still, if there is any relationship between that logic and an underlying theory of change in the food sector, it remains mysterious. The absence of theory is quickly explained by flipping the map: on the reverse side a description reads, "This map is a tool for starting conversations about how technologies can be used wisely to close important gaps in the food system." IFTF staffers tend to have an elective affinity for making tools of all kinds. They are more likely to have backgrounds in graphic design or theater than economics, and they are more likely to have studied anthropology than demography. Operating in a Silicon Valley that currently makes a fetish of an abstraction called "big data," as well as myriad forms of very real data, IFTF has recently made video games, told stories, arranged workshops, and brokered meetings between entrepreneurs and experts. IFTF staff members are more likely to speak of signals than they are to speak of data, but they know how to make those signals persuasive.

Similar maps and brochures and handouts and posters make up the visual space of IFTF events and give shape to their corresponding cognitive space. At this meeting, however, another charismatic visual element is in the mix. A man with spiky dyed hair, wearing black, stands poised at a large, poster-width run of paper that wraps around the meeting space. Equipped with a wide assortment of colorful markers, he is a graphic artist who turns the key terms and topics of our meetings into a running cartoon strip. "Graphic facilitation" seems to be the preferred official term for this service.[8] As keywords from the presentations line up in illuminated letters on the walls around us, like domesticated graffiti, it is easy to get the implication that our meeting has a certain directionality, that we are meant to see words like "conservation" and "innovation" as actors on a conceptual stage. New vectors form between them. I'm reminded of the *Batman* television show starring Adam West in the title role. When heroes punched villains, words like "Kapow!" and "Z-Zwap!" flashed up to fill the audience's vision. One concrete advantage of this meeting's aesthetic illumination is that it is easy to backtrack and remember what's been said. I ask the facilitator if it took a long time to learn his trade, and he tells me that it's an education in paying close attention. He also volunteers that because he's trying to live as an artist in the Bay Area, the extra money comes in handy. As the whole group talks about commodity foods, corn, smiling pigs, and wheat all sprout from the white paper in yellow, pink, and brown. The artist can impart visibly dynamic energy to the methane output of cow burps.

At first glance IFTF's work seems far removed from the deeper history of futurist practice in the last century, and from the ideological, philosophical, and specifically political crucibles in which that practice was formed. Yet the epistemic uncertainties of the mid-twentieth century, when the story of professional futures work effectively began, are echoed in the futurisms of the early twenty-first, including in IFTF futurism.[9] Ironically, IFTF's founders included men committed not to living with irreducible uncertainty by building the appropriate skills (the current IFTF model), but to reducing uncertainty through forecasting. Several years before IFTF started up, Olaf Helmer and Theodore Gordon, Rand Corporation employees and two of IFTF's founders, unveiled a theory of forecasting that aspired to an accuracy associated with the natural sciences. As Helmer (who had developed the method with Norman Dalkey) put it, their goal was "to deal with socioeconomic and political problems as confidently as we do with problems in physics and chemistry."[10] He was announcing the Delphi technique, named for the Greek oracle but yielding forecasts rather than pronouncements.[11]

Delphi, which is still in use in various forms, works by finding commonalities between the opinions of experts. The organizer assembles a group of experts on a given topic and asks them a set of questions, repeating the process using the same questions through a series of rounds. These questions typically concern the probability of a particular event or outcome—a nuclear attack, perhaps, to use the locus classicus of Cold War futurism as an example. After each round of interviews, a majority view is ascertained and then, at the beginning of the next round, the experts are informed. The procedure is meant to nudge experts toward convergence upon a forecast. Delphi took shape during a time when expertise, not only the kind applicable to military matters but also expertise about complex things like modern societies, enjoyed a special cachet, bolstered not only by the reputation scientists had gained during World War II, but also by postwar U.S. government spending on education and on expert advice in matters of military and social policy. More contracts for consultants to government were drawn up as the Cold War set in, and think tanks like Rand prospered. The position of Helmer and Gordon and their ilk was rising, promising to effectively eclipse the one they had set out to defeat, namely the diffuse utopianism of revolutionaries and poets. The threat of nuclear war meant preparing for both short and long futures, and it meant rewards for those who crafted the appropriate intellectual tools.

Helmer, Gordon, and their Rand colleagues represent one of two major approaches to the future that dominated expert communities in North America and Europe in the mid-twentieth century. One of these understands futurism as the pursuit of a specific state of affairs and might not balk at being called utopian. The other understands itself to be rationalist and calculating and might balk at being called ideological, imagining scientific method to be beyond bias.[12] They do not map directly onto optimism and pessimism. Helmer, for his part, was capable of great optimism about the future of technology without making utopian proposals. Certainly, the construction of a new science of prediction was a Cold War push, an effort to map out an approach to the future for the liberal West in contrast to the *prognostik* on offer from Moscow, which, while officially pursuing a particular state of affairs and informed by a theory of historical development, was nevertheless also invested in probabilities. Nor was the Cold War the only pressing context. In Western Europe and the United States, where the more rationalist brands of futurist practice took shape, the early 1960s were characterized at once by enthusiasm for technological modernity and the thoroughgoing criticism of the same. The experts in charge of Delphi concerned themselves with the possibility of nuclear war while other experts, European philosophers especially, wrote against a world in which the tools for nuclear war had come to seem necessary. In 1964 the French theologian Jacques Ellul published *The Technological Society,* in which he glossed technology as a possible means by which we could lose our humanity, sacrificing it for greater efficiency. That same year, the German social theorist Herbert Marcuse, a member of the Frankfurt School, published *One-Dimensional Man,* which damned industrial society, West and East, for its constant fabrication of artificial needs.

As forecasting spilled from the U.S. defense sector into civilian life, it did so in ways inflected by the Cold War, and it often did so under the sponsorship of a mode of economic and sociological thought on development known as "modernization theory." In 1960 the economist Walt Whitman Rostow published a work in the latter genre, *The Stages of Economic Growth: A Non-Communist Manifesto,* explicitly pitched as an alternative to Soviet models of planned development. While Rostow was modest regarding the accuracy of bird's-eye-view historical modeling, he did say that his stages (from "traditional society" to "the age of high mass-consumption") constituted "an alternative to Karl Marx's theory of modern history."[13] In 1964, the same year Helmer and Gordon published their work, the sociologist Daniel Bell was put in charge of a predictive project on the future of American society, exploring its next twenty-five years, for the American Academy of Arts and Sciences in Cambridge, Massachusetts. The result was Bell's 1968 publication *Toward the Year 2000: Work in Progress.*[14] Another extension of this work was Bell's 1973 *The Coming of Post-industrial Society: A Venture in Social Forecasting.*[15] As Nils Gilman points out, one of the central political problems for liberal modernization theorists was the status of the Third World and the contest for hearts and minds in developing countries, which had to be secured at all costs lest they fall under the Soviets' sway.[16] But one of the intellectual problems was that, unlike their more utopian counterparts, the modernization theorists had no picture of an end state for progress that they could offer. In other words, they believed in a form of progress resembling the graph of an asymptotic formula, always approaching but never equaling numerical value one. This is a key difference between a futurism whose ideal would be prediction and a futurism that, because it is utopian, can tell you the taxonomic nomenclature of every tree in its Eden, and perhaps a pharmacology of the effects of its fruit. These two futurisms feel very different when you inhabit them, and they call for different kinds of work in the present. Contemporary IFTF futurism matches neither, but sometimes notes of each of these futurisms can be detected in IFTF scenarios.

The histories of capitalist democracy and of socialist, or more precisely Marxist, thought do echo through these kinds of futurisms, but matters are far more complicated than prediction aligning with capitalist democracy and utopianism aligning with socialism. There are capitalists who propose utopias produced through the mechanisms of the market, and socialists concerned with local issues of strategy, who understand that, despite the element of Hegelianism in Marx's thought, Marxist history is scarcely the unfolding of reason over time, its progress both teleological and organic. Olaf Helmer hoped for the arrival of "constructive utopians," showing that he had no fear of the word. Some fantasies about the future of cultured meat underscore the point that views on the future rarely align with politics or method; many vegans and animal welfare advocates with whom I spoke place their trust in market economics to effect social change, but they also believe we are moving toward a particular and knowable end state, a utopia in which animals are no longer harmed. Futurist conversations are notable for their remixing of ideological positions, and they are better places to find eclecticism than consistency. But a more historically specific reading is also available: a generation after the fall of the Soviet Union, when the triumph of the market seems (to many) certain, capitalist techno-utopias happily coexist with the intellectual tools for reducing uncertainty, perhaps because such positions scarcely threaten one another. Teleological tendencies are not a sign of Marxist thinking, merely another position available in the marketplace of ideas in a free-market world.

We're still at the card tables, looking at the bugs. After learning a little more about the conditions in which mealworms like to grow, and the things you can bake with mealworm flour, I turn my head. The main event hall at IFTF and the rooms adjoining it are decorated with what our hosts call "artifacts from the future." These three-dimensional mock-ups or flat images, created by IFTF designers, draw our attention throughout the weekend. Keepsakes from other projects, these include gas masks for a future of toxic air and pictures of futures in which synthetic biology grants us an unprecedented level of control over the life around us, bacteria producing biofuels while genetically modified trees sequester as much carbon as we can burn, saving the air. My personal favorite, created by Sarah Smith, a member of IFTF's food team, is called the "Meat Counter of Tomorrow." A depiction of a butcher shop's glass-fronted display case in 2023, the image is crowded with familiar-looking cuts. The difference hides in their labels: "grassland rehab grazed beef top loin steak—boneless," "grade A road kill—venison chuck," "in vitro lab grown pork shoulder." This artifact's overt point is that if we are nimble and creative our meat supply can become more sustainable. More subtly, it suggests that conventional industrial meat is unsustainable and that in the very near future we will have to chew what we previously eschewed, either by tapping previously untapped sources or by exploiting new technical means. The "artifact" is one snapshot out of a future storyline, plunking the viewer down in medias res. No exposition is provided to explain just how we got to this particular future, but it is not tough to imagine how each cut in this "Meat Counter" might have come about.

Encouraging the viewer to do this storytelling work is precisely the point of such "artifacts from the future." I can fit a tale to each cut of meat. In the United States, rural drivers pass pieces of roadkill regularly. These are animals that have become meat through unhappy accident, or as one writer called it, "manna from minivans."[17] Although, in some states, food banks have embraced what is also called "vehicular venison"—in Alaska all roadkill is technically owned by the state and used to feed the needy—in general it is an underused resource. "Grassland rehab grazed" refers to what is sometimes termed "holistic grasslands management," promoted by the biologist and ecologist Allan Savory. It aims at undoing some of the desertification of the earth's grasslands, often thought to be the result of overgrazing, through careful husbandry and selective measures of "rewilding"[18] that could (in theory) enable environmentalists to eat beef, generally acknowledged to be the most environmentally damaging meat, with a clean conscience. The notion that we'll have "in vitro lab grown pork shoulder" by 2023 makes me grin a little. It seems wildly optimistic to expect such a structurally complex piece of tissue in such a short time, even though we've just seen the first lab-grown hamburger. That Smith has included it here is probably a sign of the increasing public visibility of lab-grown meat, rather than evidence that IFTF has a pipeline to a secret advanced meat lab.

If roadkill, grasslands rehabilitation, and in vitro techniques are three futures for meat imagined in the "Meat Counter," they are also part of a larger story about how we might continue to eat meat in the future, albeit in a reformist fashion. Smith's "Meat Counter" has left industrial animal agriculture behind, but it is still capable of satisfying appetites much like our present-day ones. This is a future in which production or supply has changed, and perhaps our sense of meat's ontology has shifted as well, but meat's central place in our diets has not. In this way the "Meat Counter" harmonizes, intentionally or not, with a dominant assumption encountered again and again in the history of the future of food: meat is a measure of plenty, and a set part of a normal and healthy diet, which in turn might be used to determine how large a population an area of agricultural land will support.

Meat enjoys a disproportionately large place in futurist visions of food production, and it is possible to see Smith's "Meat Counter" as a wry smirk about this.[19] For over two hundred years in Europe and North America, conversations about the future of food have been driven forward by the fear of not having enough food to go around. But meat isn't efficient. To create meat, we sequester a great deal of plant food (not to mention water and, figuratively speaking, land) in a relatively small amount of animal food. In Book II of Plato's *Republic*, Socrates, in dialogue with Glaucon, frames this as an international relations problem thousands of years before the rise of the modern nation-state: the land requirement for raising livestock for meat, which he understands as a luxury food, leads to the constant need for more and more territory. Meat meant wars of expansion. Frances Moore Lappé's book *Diet for a Small Planet* (1971) did much to publicize the ratios between animal feed and meat, for various species, offering the dismally poor 21.4 to 1 for cows. Such figures support the notion that bovines are, in fact, machines for generating inefficiency in the food system, even when they are not busy being a cause of bellicosity.

I pause for another grasshopper, which is tasty and very much not a part of the ecological footprint of affluent societies. I chew on the issue of the inefficiency of cows. The assumption that meat (and especially the Anglo-American beefsteak) measures plenty is widespread within think-tank food futurism, a genre significantly older than the versions of professional futurism that follow Rand. According to Warren Belasco, Anglo-American thinkers have long featured meat in their projections for a desirable standard of living, and in their claims about demographic limits for human populations. Beginning at the end of the eighteenth century, a preference for meat made its way from British pens into policy and influenced the lives of people living in British colonies, particularly in India and Ireland.

Perhaps the most influential pen was held by Thomas Robert Malthus (1766–1834), cleric of the Church of England, father of the field known as political economy, and eventually a professor at the British East India Company's college. Malthus was well aware of meat's inefficiency compared to plant foods, but he nevertheless presumed that meat was so highly prized that it should influence policy on matters related to population. If population growth seemed to lead inexorably to the use of more farmland for plant foods and less for meat, this was a reason to limit reproduction, rather than to eat plants in place of animals.[20] Malthus seems not to have based his position on meat on personal taste. He saw meat as a kind of cultural necessity others would not readily give up, something that kept up eaters' morale, but he understood that rising beef prices meant that meat might be limited to those wealthier people who could afford it. His views on population and population control, which would become the doctrine of Malthusianism, thus presupposed omnivory and were edged with worry about meat.

In his most influential work, the 1798 *Essay on the Principle of Population*, Malthus proposed a theory of the relationship between human appetites (gustatory and sexual) and our powers of production. Looking to late eighteenth-century schemes for improving the yields of the British countryside, he understood agriculture as subject to improvement by technology and technique—to a degree.[21] But the appetites were greater. He judged that populations tend to increase geometrically (i.e., exponentially), whereas food production only tends to increase arithmetically (by addition). The result is that we can expect food production to regularly fall short of population growth, placing the poorest among us at risk of malnutrition or starvation. This pessimistic picture was joined to a dismal view of the poor themselves. Malthus characterized them as unable to constrain their appetites. The most influential late twentieth-century Malthusian in the Anglophone world was Paul Ehrlich, who is still active as of this early twenty-first-century writing. With his wife, Anne, he authored *The Population Bomb* (1968), which foresaw widespread famine as early as the 1970s and proposed draconian measures ranging from forced birth control and lotteries for the right to procreate (in the developed world) to food aid doctored with antifertility drugs (for the developing world).[22] It was through the Ehrlichs' work that many at this IFTF conference first encountered Malthus's thought, myself included. Some of us, I learn through conversation, are even aware of a famous $10,000 bet Paul Ehrlich once lost to the more optimistic economist Julian Simon. Simon, a celebrant of the powers of growth, bet that the real cost of a group of five raw materials, namely chromium, copper, nickel, tin, and tungsten, would not rise between 1980 and 1990. The absence of widespread famine (in the developed world) after *The Population Bomb* seems to have settled widespread opinion against Ehrlich, and the Ehrlich-Simon wager has reassured celebrants of economic growth that resource extraction and production will continue to meet rising demand. Yet Ehrlich himself has not recanted his views in any meaningful way,[23] and the early twenty-first century has many convinced Malthusians, whose calculations involve not individual nations but the human "carrying capacity" of the globe itself.

Malthus had contemporary critics in the circle of the vegetarian utopian socialist William Godwin. They counseled a diet of legumes and grains and worked from premises that may seem strange from an early twenty-first-century perch, namely that a plant-based diet would support more humans and that a larger population would somehow increase the sum of human happiness. That goal alone—maximizing human happiness as opposed to merely ensuring human survival—reminds us that the future has looked very different to different eyes. Belasco, offering an overarching typology of the competing positions on the future of food, juxtaposes Malthusians and Godwin-inspired "egalitarians" against a third group, who in fact would exert the most influence on the shape of modern food systems in the developed world. These are the "cornucopians," a fitting name for anyone who believes that production, not merely of plants but of animal flesh too, can keep pace with population growth. Those trying to reconstruct the history of policy debate might turn to the Marquis de Condorcet as an eighteenth-century cornucopian thinker roughly contemporary with Malthus, but "cornucopianism" is a less self-conscious school of thought than Malthusianism. It is also more widespread, because it includes not only policy experts but also entrepreneurs who are effectively cornucopian in their thinking, whether or not they recognize the label. As historian Fredrik Albritton Jonsson puts it, "these two forces have been feeding on each other, generating rival forecasts of technological development," not to mention rival accounts of the limits of what technology can provide.[24] Albritton Jonsson notes that in 1817, David Ricardo expressed his cornucopian views in the straightforward terms of political economy. If Malthus thought that soil was only finitely exploitable, Ricardo suggested that labor and invested capital could improve apparently inferior soil. The extension of Ricardo's principle is that we could exhaust a particular natural resource in a given location and then move on, finding new resources susceptible to our ingenuity, allowing us to extract ever more value. Ricardo's view mirrors the apparent economic lessons of the industrial revolution.[25] And the notion of continuing economic growth by simply shifting natural substrates, from fossil fuels to sunlight via solar panels for example, can also be found within Ricardo.[26] Also foreshadowed, albeit in a weirder and more roundabout way, is the contemporary notion that economic growth might proceed fully decoupled from environmental resources, through the information economy.[27]

"Nature," Condorcet wrote, "has set no limit to the realization of our hopes." Many of the developments we've learned about during this weekend at IFTF are cornucopian in the sense that they approach problems of food supply by changing the substrate of subsistence. Sometimes, as in the case of the fantastical "matterstream," they propose a new type of production that would completely change the system of territorial control, resource extraction, material purification, and fabrication established during Europe's Industrial Revolution. Notably, Godwin's circle, and other egalitarians (and socialists) after them, would often smile on the possibility of agricultural improvements, including technological ones, keeping pace with population growth; Friedrich Engels, Karl Marx's coauthor and a prominent critic of Malthus, certainly did. Socialists, like capitalists, have long entertained fantasies of establishing a kind of technological second nature.

The current cornucopian argument, usually advanced by staunch believers in the free market, is that resource depletion and climate change only demand a slight tempering of the confidence in growth that has animated economic thought since World War II. Human ingenuity will out; we will adapt and thrive, perhaps beyond the limits imposed by fossil fuel dependency. From this cornucopian perspective, the continued growth of industrial meat production merely faces a series of challenges, environmental and perhaps ethical. The appeal of laboratory-grown meat is that it promises to make those challenges temporary. The bioreactor could become a new source of natural resources, or a new artificial frontier within what Albritton Jonsson terms the "economic doctrine of indefinite substitution."[28] Not all advocates of cultured meat are cornucopians, I will learn in the course of my research, but many deserve the label.

From the mid-nineteenth century through the twentieth century, increasingly concentrated and efficient industrial animal agriculture made meat the paradigmatic food of cornucopianism in the Western world. Industrial animal agriculture has been part of what Will Steffen, Paul Crutzen, and John McNeil call the "Great Acceleration" of growth in affluent societies between 1950 and 2000, which more than doubled the world's population, from 2.5 to 6 billion, and increased the carbon dioxide in our air by a third in that same time frame.[29] The effort to shift eaters to other forms of protein (such as insects) that act not merely as alternatives to meat but as explicit substitutes for it is, from this vantage, deeply suggestive. This effort recalls Ricardo and his vision of shifting from one piece of soil to the next as we exhaust the land. It means that many of our forms of food futurism are shaped by an ideology of nimble and continued growth even as they try to ameliorate growth's worst side effects. The "Meat Counter of Tomorrow," for all its creative charm, also maintains the continuity of meat consumption patterns begun during the mid-twentieth century, the era when cheap meat became part of the infrastructure of our lives. The "Meat Counter" posits that we are attached enough to familiar-looking pieces of meat that we will exploit any resource that can provide them.

The "Meat Counter" also implies a society that has lost or willingly relinquished conventional industrial-scale meat production. Presumably, in such a society, the virtues associated with carnivory (in its new form) would include thrift, resourcefulness, and flexibility in the face of climate change. But such a society might also have a transformed view of what it means to be modern. Consider the words of the sociologist and modernization theorist Edward Shils, delivered in Dobbs Ferry, New York, as part of a keynote address on the predicaments of "new states" recently created in the shuffle of post–World War II decolonization:

> No country could be modern without being economically advanced or progressive. To be advanced economically means to have an economy based on modern technology, to be industrialized and to have a high standard of living. All this requires planning and the employment of economists and statisticians, conducting surveys to control the rates of savings and investments, the construction of new factories, the building of roads and harbors, the development of railways, irrigation schemes, fertilizer production, agricultural research, forestry research, ceramics research, and research of fuel utilization. "Modern" means being western without the onus of following the West. It is the model of the West detached in some way from its geographical origins and locus.[30]

The meat of such a world must be cheap, and industrially produced at great scale. If it is easy to see how industrially produced cultured meat might comport with Shils's vision, it is hard to see how roadkill would fit in. Grazing cattle to revitalize grassland also fits awkwardly here. It implies a close attention to geographic origins rather than the global rootlessness that Shils implied, the picking up and plopping down of production strategies without much heed for the land.

The "Meat Counter of Tomorrow" tells multiple stories: the presence of cultured meat suggests technology's continued rise, while the other meats suggest old technologies of meat production falling by the wayside, replaced by different kinds of creative repurposing. It implies the failure of an older model of modernization, which happens to be the one that Shils described. The hamburger is the meaty mirror of Shilsian modernity.

While the IFTF staff themselves endorse no particular ideology, at least not explicitly, much of what we see during "Seeds of Disruption" has been pulled in from Silicon Valley. Quite literally: we hear from a biochemist who develops plant-based and egg-free mayonnaise for a company called Hampton Creek Foods (which will later become involved in cultured meat research); a representative from a company that manufactures sous vide devices for home cooks demonstrates cookery in a carefully calibrated water bath. Then it's time for a field trip. The IFTF team leads us on a two-block walk to a location of the Whole Foods grocery chain, which would be considered upscale in most parts of the country but in Palo Alto counts as a neighborhood market. "As part of the process," the schedule reads, "participants will be asked to purchase a food product that is on the verge of being disrupted in the next decade." I walk the aisles and chat with fellow workshop participants and IFTF staff, asking, "What do we mean by 'disruption'?" In the media discourse surrounding Silicon Valley, the term has recently become a target for criticism as a piece of obfuscating jargon. A term first popularized in business school literature, "disruption" seems to have become more signifier than signified. When an existing industry becomes disrupted (as cellular agriculture might disrupt industrial animal agriculture), younger companies swarm in, their technologies offering advantages the old ones lacked. Cultured meat is thought to be "disruptive" because it is "innovative." The relationship between the two terms is almost reflexive, containing a two-word story that many people hold to be true.

I try to think through this as we pause in the toiletries aisle. The words are tired, overused, and perhaps over-attacked; in a few years they will sound quaint. "Disruption" and "innovation," and especially both terms used in combination, suggest that existing areas of business (or "spaces," to use that territorial metaphor) succumb primarily because their modus vivendi falls out of step with the pace of change. Less established, more nimble players arrive and keep pace quite nicely. But if the whole point of disruption is that the center cannot hold, why should disruption not fall prey to time too, particularly as the vogue for venture-backed small businesses wanes with available venture capital? We stop and examine colorful tubes of toothpaste and ask if some innovation in dentistry might eliminate our need for them. Some commentators have observed that the notion of disruption is an alternative to the idea of gradual progress, which is evidently staid, old-fashioned, obsolescent. Others suggest that the strange thing about the innovation-disruption dyad is that it makes every success story into a tabula rasa, so that the gaming pieces that once stood on the table now lie scattered on the floor. Innovation-disruption eliminates continuity and ignores its value. And this brings us around to the fundamental dilemma of innovation in food systems, from the standpoint of the members of the managerial class visiting Palo Alto this weekend in 2013: from agriculture to consumption, the food industry relies on continuity and reliability. Yet it is buffeted by the constant and slower-moving disruptions of social and environmental change, and by the much-discussed, fast-moving, but untested disruptions created by new companies and new products. The attendees at "Seeds of Disruption" are mostly representatives of the kinds of companies that worry about being disrupted. Before my research has concluded, several years down the road, major food companies will be making exploratory investments in cultured meat technology, perhaps driven by that same worry.

I am not the first observer who, when confronted with the notion of disruptive innovation, reaches for an apposite passage from Marx's *Communist Manifesto:*

> The bourgeoisie cannot exist without constantly revolutionizing the instruments of production, and with them the relations of production, and with them all the relations of society. . . . Constant revolutionizing of production, uninterrupted disturbance of all social relations, everlasting uncertainty and agitation, distinguish the bourgeois epoch from all earlier ones. All fixed, fast-frozen relations, with their train of ancient and venerable prejudices and opinions, are swept away, all new-formed ones become antiquated before they can ossify.[31]

Marx's argument was that an ultimate revolution that brings us beyond capitalism would spring from within a capitalist world, stemming from the bourgeoisie's constant impulse to reinvent itself.[32] The resulting phenomenon—"All that is solid melts into air"—has implications that run deeper than particular political and economic answers, including communism itself. It means the possibility, indeed the near certainty, of reinvention without reason, of new ideas unmotivated by real human needs.

My team picks up packages of nuts, pieces of fruit, envelopes of beef jerky, and squeeze bottles of ketchup, all of which we have decided are ripe for change. There are, in fact, reasons to not want food to be disrupted at all, and to find ways, including technological ones, to preserve our food system as best we can. I think about water, one of the natural resources expected to dwindle in availability in decades to come; I look at a package of California almonds, that water-intensive crop. It would be easy to grab a package of conventional hamburger meat and argue that cultured meat will "disrupt" it, but I don't feel like taking responsibility for that future story line just yet.

NINE

Prometheus

Summer 2014. After a day spent walking around Cork, Ireland, Ryan Pandya emails me a quotation from the early twentieth-century geneticist, and outspoken socialist, J. B. S. Haldane. It's excerpted from Haldane's book *Daedalus* (1923), which began as a lecture delivered to a Cambridge University club called the Heretics' Society. The quotation begins, "The chemical or physical inventor is always a Prometheus. There is no great invention, from fire to flying, which has not been hailed as an insult to some god." Haldane continues in a fashion congenial to Pandya's own criticisms of industrial dairying, which we spent much of today discussing, as we strolled across the green campus of University College Cork. Haldane suggested that for humans to take milk from the udder is "indecent" to the cow, a kind of violation of the "intimate and almost sacramental bond between mother and child." Pandya's goal is to replace the cow, in the process of milk production, with the methods of cellular agriculture. This is why Pandya and his cofounder, Perumal Gandhi, have created their company, Muufri, which at this moment is still in its earliest phases, hosted within an incubator program that is using laboratory space at the university. They don't fit cows into a sacred register as Haldane did. They talk about the structural violence inherent in industrial dairying and implicitly elevate the moral status of animals by recognizing their suffering.

Don't skip blithely past Haldane's invocation of Prometheus. Like many other writers who associate Prometheus's gift of stolen fire with the beginnings of technology and civilization, Haldane casts Prometheus as an inventor. But Prometheus did not create fire. He stole it. Nor, in some versions of the tale, was it his first theft from the gods. Though his theft of fire overshadows it, Prometheus first stole meat.

In his *Theogony,* Hesiod describes Prometheus intervening in Man's very first sacrifice to the gods ("Man's" because humanity was then only male, single-sexed). The sacrifice was an ox. Prometheus took a pile of the best cuts and laid the stomach on top to serve as an offal-ish concealment. The other pile was nothing but bones. Prometheus layered shiny fat over them, hiding their dull whiteness. He asked Zeus to choose his favored portion, and while Hesiod implies that Zeus recognized the deception, the father of the gods nevertheless willingly and knowingly selected the bones, and then Prometheus gave the meat to Man. In retribution for the attempted trickery, Zeus kept fire out of human hands. The relationship between this first theft and Prometheus's second, more famous one, is ambiguous: was it Zeus's interdiction of Man's access to fire that made Prometheus's second theft necessary?[1] Or was it simple culinary logic that demanded that a theft of meat should be followed by a theft of fire with which to cook it?

The main story scarcely needs recounting.[2] Prometheus stole fire from the home of the gods on Olympus, and he gave it to Man. In many retellings, such as Haldane's, the gift of fire was also the gift of civilization, because fire meant controlled access to a source of heat and light, greatly expanding the scope of human activity beyond the day and into the dark of the evening, and extending the possibilities of human culture beyond the diurnal fulfillment of physical needs. Zeus's punishment of Prometheus is as well known as the story of the theft. The titan was chained to a rock, where an eagle would land on him, tear open his belly, and eat his liver. This organ would then regenerate, enabling the eagle to repeat the performance the next day. Though in some tellings of the tale Herakles ultimately rescued Prometheus, the titan's punishment was originally intended to be eternal. It symbolically linked "living" fire and living flesh, because of Prometheus's capacity not merely to heal but to fully regrow his own tissues. To reverse a phrase used by historian Hillel Schwartz, myth itself is invested with the rich ambivalence of biology.[3]

In Hesiod's version of the story, the theft of fire also transformed human biology at the reproductive root. Man was punished along with Prometheus, through the creation of Woman, who is described in the *Theogony* as a distraction and a possible economic burden for her potential spouse, though another obvious point peaks around the edges of this rather misogynistic story: Woman also became the key to the perpetuation of the species via sexual reproduction. Perhaps not life in the most basic sense, but sexual, reproductive life as we know it, thus comes from the theft of fire. In his book *The Psychoanalysis of Fire,* the philosopher of science Gaston Bachelard made the poetic gesture of linking fire to life as twin explanatory principles that function at different speeds: "If all that changes slowly may be explained by life," he wrote, "all that changes quickly is explained by fire."[4] Bachelard is gnomic and supplies no argument, but he seems to be invoking the distinction between nature and culture, or between the pace of change when humans are not on the scene and the pace of change after humans and their artifice arrive. The desire to control fire, Bachelard further claimed, was part of a "Prometheus complex," the intellectual will to know more than our parents and teachers. That complex sparks up when a child's parents forbid her to play with matches.

No wonder the story of Prometheus has been so generative in its retellings.[5] It is a story of origins and regenerations layered upon each other: meat, fire, Woman, not to mention human sexual reproduction, all from a primal transgression. The human family begins with a betrayal of the gods' proper worship. Suspended on a rock above all this is the eternal liver of a titan, those deities who preceded Zeus's own heavily godded Olympian pantheon. Small wonder that Prometheus is also sometimes described as the creator of the human race itself. Or, as in Plato's account of things, even if the gods created humanity it was Promethean fire that carried humans beyond an animal existence. Prometheus's brother, the titan Epimetheus, had already given all manner of advantages (teeth, claws, baleen, feathers, armor) to the nonhuman animals. Prometheus's name can be translated as "forethought" and Epimetheus's as "hindsight" (and Epimetheus is often described as Prometheus's less thoughtful twin). In a case of inadvertent foreshadowing, Haldane's *Daedalus* included a reflection on the potential of in vitro techniques for human fertilization and reproduction, written almost sixty years before the first babies were conceived through in vitro fertilization techniques. This reflection effectively revitalized Hesiod's implied link between Prometheus's deeds and the future of human reproduction, but it also made Prometheus into an appropriate patron deity for cellular agriculture, which makes its own use of in vitro techniques—doubly appropriate when we consider that in Aeschylus's version of the Prometheus tale, the titan is credited with teaching humans to domesticate and harness animals. Why should he not also unharness them, as it were, by freeing them from industrial agriculture?

It is a neat further coincidence that Prometheus's tale was invoked in the vegetarian literature of nineteenth-century Britain. In 1813, nine years before he drowned off the Ligurian coast, the vegetarian poet Percy Bysshe Shelley published a short essay entitled "A Vindication of Natural Diet." This was the occasion for his well-known quip, "The monopolizing eater of animal flesh [destroys] his constitution by devouring an acre at a meal." Shelley cited a reading of the Prometheus myth offered by Dr. John Frank Newton, published in Newton's *The Return to Nature, or, A Defence of the Vegetable Regimen* (1811). In Prometheus's double theft of meat and fire Newton saw the start of human carnivory, and thus he located the very beginnings of our decline in our very origin, a notably Biblical rendering of Greek myth. Disease, precarity of existence, and the end of a previously lifelong youth—all were the fruits of cooking and consuming animal flesh, according to Newton and Shelley. In Shelley's essay this citation to Prometheus in turn demands reflection on a paradoxical problem: how to preserve the gains of civilization Prometheus won for us, while rejecting "the evils of the system, which is now interwoven with the fibres of our being." Perhaps carnivory need not be our permanent condition. One could perhaps return to what Shelley called "nature" (or natural diet) without abandoning the culture that the titan lit.[6]

TEN

—————

Memento

Discourse has broken down and the chefs seem drunk and surly, so I get up, walk to the back of the room, and open a can of beer. Life is short all around us. More accurately, a series of identical posters declaring that life is short hang from the walls at either side of the room, positioned at regular intervals like heraldic flags in the great hall of a castle. They might produce a stereoscopic vision of mortality if only our eyes were positioned on the sides of our heads, like fish eyes. While "Life is short" is the main message, in larger letters, the posters are crowded with other notions: "Live your dream and share your passion." "If you are looking for the love of your life, stop; they will be waiting for you when you start doing things you love." There's even an injunction meant to ward off an unwanted anthropologist or literary critic: "Stop overanalyzing." We're in the middle of a failing conversation at a place called Holstee, which is both a design shop in Brooklyn and a business built around the messages on the posters—also available on coffee mugs, greeting cards, and seemingly everything but toilet paper. It's a clean, modern space, one in which neutral-colored walls make room for Holstee's revelation of life's brevity and insistence that we realize all of life's possibilities in a reassuringly responsive universe. According to Holstee's slogans, the ultimate problem is neither the inhumanity of human to human nor the ground-level "meaninglessness of it all," but the fact that the clock runs out. It is difficult not to think of a passage from Arthur Schopenhauer's *The World as Will and Representation* (1818), in which the post-Idealist German philosopher writes:

> The completed course of life upon which the dying man looks back has an effect upon the whole will that objectifies itself in this perishing individuality, analogous to that which a motive exercises upon the conduct of the man.

It gives it a new direction, which accordingly is the moral and essential result of the life.[1]

Incidentally, the universe seems entirely indifferent to my own feelings of skepticism regarding Holstee's account of how life's purpose springs from life's end. Nevertheless, Holstee's clean-lined and well-kerned epiphanies (the design is nice) have set the stage for tonight's event, an evening of casual conversations about cultured meat and its potential benefits and risks. The chefs have become a little agitated about cultured meat, though they aren't the only ones. When asked if he would consider cooking cultured meat in his restaurant, one of them laughed "No!" and encompassed in spittle the tone of the evening's conversation.

The issue ostensibly on the table is the "naturalness" of lab-grown meat in the eyes of its potential cooks and consumers. The conversation and its failure matter not because we've been getting to the bottom of things, but because there is something instructive about our inability to find a purchase on the deeper issues at hand. Rather than sift through the potential meanings of the words "natural" and "unnatural," perhaps examining how they relate to both industrial methods of agriculture and small-scale and organic methods, we have mostly yelled. Our ideological agendas have made it impossible to come to common terms about the goals of agriculture and food production, let alone the meaning of the meat some of us eat.

To adapt the terms of the anthropologist Heather Paxson, the critical anti–cultured meat claim in this conversation is that cultured meat would exist outside a "moral ecology of production."[2] Such a moral ecology would be a virtuous circle that encloses producers, consumers, and natural resources, a circle sustainable at the timescale of generations rather than product cycles. In other words, in this conversation cultured meat is being compared not to conventional meat produced at an industrial scale, but to small-scale and environmentally responsible animal farming. As Rachel Laudan observes, the distinction between "natural" and "unnatural" foods is evocative, but it is invoked more often for its rhetorical power rather than for its explanatory usefulness.[3] It is often used to support what Laudan terms "culinary Luddism," namely the rejection of what she calls "culinary Modernism," and the two terms are nearly self-explanatory; there is the smashing of the machines and the celebration of the machines, which can press tortillas, mill flour, conch chocolate, and roast coffee. Luddism, which would find the notion of a "moral ecology" entirely congenial, "involves more than just taste"—"it is a moral and political crusade."[4] Different levels and forms of culinary Luddism can be found in the books and articles of food gurus, in cookbooks, in the graphic design of restaurant menus decorated not with illustrations of industrial flour mills but with vines, pitchforks, and cast iron skillets. There is, it must be said, a market for Luddism, much as there is a market for modernism.

Tonight's event, in which Luddism (which might more charitably be called neo-agrarianism) and modernism clash via the exemplary case of laboratory-grown muscle tissue, is called "Food Fight: An In-Depth Look at Cultured Meat." At this moment, as I grab my beer, we're several hours in, on the unseasonably warm, rainy evening of November 13, 2014. The event has been organized by a group of journalists from three organizations: the very well-established science and technology journal *Popular Science;* Climate Confidential, a team of reporters who run a website about climate issues, agriculture, and technology; and *Modern Farmer,* a magazine that addresses issues in food and agriculture, with a sensibility balanced between hip urbanity and practical farm experience. By browsing *Modern Farmer*'s website, you can watch the pastoral adventures of ruminants through a remote "GoatCam." My guess is that Climate Confidential arranged my invitation and made me a speaker on the first of the several panels that make up tonight's event. I've been chatting with them off and on for a few months, sharing my initial research on cultured meat, aware that such friendly sharing sometimes yields invitations to events like this one.

Holstee doesn't have a direct voice at "Food Fight" (just those looming posters), but it has a few ideas of its own about food. Holstee's designers have produced a poster, called "Food Rules," inspired by a classic American World War I poster that offered guidelines for food practice on the home front, including such advice as "Buy Local Foods." This directive from the second decade of the twentieth century harmonizes with one kind of food wisdom common in the early twenty-first century, namely that minimizing the length of transportation chains between producer and consumer reduces the carbon footprint of our food system.[5] Those transportation chains, originally a fruit of nineteenth-century urbanization, have hooked city and country together, and transformed huge swaths of North American territory and biomass to suit the needs of an urban and urbanizing population. William Cronon has used the phrase "annihilating space" to describe the meat industry's modernization. This double entendre refers both to the stockyards of nineteenth-century Chicago (one of the cradles of the modern meat industry) and to the way supply chains for meat literally annihilated the physical distance between an animal's gestation and birth, its life processes, its subsequent killing and butchering, and its eventual consumption.[6] Supply chains are as much engines of modernization as industrial assembly lines, and according to the so-called locavore perspective on the reform of industrial food, supply chains must be shortened. Locavores, as the name suggests, believe in eating locally. What follows from this (not uncontested) wisdom is that we should encourage local farming, and perhaps promote agriculture within urban centers, which promises to reduce the distance between farmers and consumers. Urban farming, either via hydroponics or stacked, vertical farms in skyscrapers, or simply in urban warehouses or lots, is a favorite topic at conferences on the future of food.[7]

Holstee's neo-agrarian sentiments present a striking counterpoint to the defining hydrological feature of the design firm's neighborhood. Holstee's offices are in Gowanus, Brooklyn, just a short walk from the Gowanus Canal. Built in the nineteenth century, the canal was used as a transport route for Brooklyn's industrial plants, including paper mills and tanneries. The warehouses around Holstee's offices bear the faded letters of factory names, reminders of an erstwhile regime of fabrication and consumption and the export of goods made in New York City. Before the canal was dug, the swampy lowlands of Gowanus received the sewage flowing downhill from the higher-elevation and wealthier neighborhood of Park Slope, which lies to the east; Carroll Gardens lies to the west, Red Hook to the south, and Boerum Hill is north. Throughout the twentieth century, industrial waste would combine with sewage, particularly in times of heavy rain when Brooklyn's sewers overflowed, giving Gowanus a reputation for pollution it has maintained ever since. In 2010 the Environmental Protection Agency gave the canal Superfund status, and its 1.8-mile run has been recognized as one of the most polluted waterways in the United States.[8]

Nor is there anything agrarian, neo- or otherwise, about the ironic celebrations of pollution visible in certain displays of Gowanus-area culture, celebrations that seem to have followed behind the real estate boom in this part of town, a boom that arrived late by Brooklyn standards. The Gowanus Yacht Club, a bar, would not bear its ironic name if there were any yachting to be done on the canal's short, bridge-dotted span. The name of the Lavender Lake Bar refers to a nineteenth-century description of the canal's appearance. If the top was purplish in those years, the sludgy bottom was called "black mayonnaise." Some speculate that those celebrations result from a specific tension between the gentrification of the surrounding neighborhoods in Brooklyn and the comparative lag of Gowanus's gentrification, making Gowanus a borderline-chic, postindustrial preserve, effluence and affluence together at last. Holstee's designers craft their messages of mortal fulfillment in a neighborhood whose community members once invited, and then celebrated, the Environmental Protection Agency's proposal to designate the canal a Superfund site. They knew that status would help protect their neighborhood from real estate developers. Now a branch of the upmarket Whole Foods supermarket chain stands nearby. I used to live several blocks away, in Carroll Gardens. I left the neighborhood before Whole Foods opened, but I can remember the debates about what the supermarket's arrival would mean for the area. It wasn't a question of whether the neighborhood had gentrified, merely of what point along the gentrification curve we had reached. And of course there was the question of whether families that had lived in the area for generations would be able to remain. During my time in Brooklyn, visitors who wished to explore the natural history of the canal could do so at a few exhibits in a small museum and art gallery called Proteus Gowanus, not very far from where Holstee is now, located in a brick building just off the canal.[9]

When I arrived at Holstee the office's large open hall was still set up with rows of desks. The event organizers greeted speakers and guests as we trickled in. The day's work had just ended and the evening's was beginning. Gradually the desks disappeared, replaced by rows of chairs facing a platform at one end of the room, an impromptu stage for the discussants. In a side room, a reporter pointed a video camera at Andras Forgacs, cofounder and CEO of Modern Meadow, recording an interview about the company's efforts to create meat (efforts that would end later in 2014; Modern Meadow would, in later years, help establish a separate company that aims to produce meat) and a leather-inspired material made from collagen using tissue culture techniques. Having met Forgacs before, I knew him as an enthusiastic spokesman on behalf of emerging biotechnologies. His role in tonight's discussion was to represent cultured meat, and he would do this against significant opposition, much of it coming from celebrants of a set of neo-agrarian values for food that are close to those represented by Holstee's little food poster: locality, organic-ness, and authenticity, often invoked as constituent parts of a moral ecology of production, consumption, and land stewardship.[10] These are, as Laudan points out, also very modern things (as opposed to traditional things) to want food to be, and have more to do with a faux nostalgia than with a historically acute sense of what and how our ancestors ate. I am not the first to observe that in the early twenty-first century, in the richer parts of the developed world, food acquired a new status as a site where these two meanings of "value" seemed to merge, the behaviors of capitalism (producing, transporting, selling, buying) seemingly serving moral purposes as well as economic ones. Work in artisanal food production has, similarly, taken on the aspect of a vocation, in something like the process Max Weber discovered in what he called the "Protestant ethic."[11] The deepest failure in this evening's conversation was not our inability to agree on a definition for "natural," but our inability to see our own deep investments in ideological purity when it comes to the value of artisanal food production compared to that of industrial-scale food infrastructure.

Most of the audience members who trickled in were young, well dressed, and white, which did not surprise me; conversations about cultured meat tend to draw in audiences that are younger, well educated, and relatively affluent, although often diverse in terms of racial and religious background. Our hosts welcomed the audience and introduced the panelists, and the conversation began. Up first, Forgacs and I took the stage and I was momentarily distracted by the words printed on his jacket: "Solve for X." This is the name of a series of talks organized by Google in which guest speakers propose solutions, often technological, for grand global challenges. "Solve for climate change," perhaps, or "Solve for cruelty to farm animals." Forgacs must have spoken in the series. Such jackets or t-shirts are not uncommon swag for speakers at conferences in Silicon Valley, and they send a highly specific social signal, a little like the letterman jackets once commonly worn by American high school and college athletes. Forgacs is young, but Modern Meadow is not his first company. He and his father, Gabor Forgacs—a scientist whose first field was theoretical physics and whose area of expertise is biological physics—previously founded Organovo, whose goal is to print human tissue suitable for testing drugs.

In his introductory remarks, Forgacs described Modern Meadow and its mission as of this point in 2014. The inspiration for Modern Meadow came from Organovo's work with tissue engineering and 3D printing techniques. If we can print human tissues, why not print nonhuman animal tissues for use as clothing or food? In a talk given at the TED (Technology, Entertainment, Design) Global conference in June 2013 (about two months before Mark Post's cultured beef media event), Forgacs presented Modern Meadow's goals. He began by suggesting that, by the middle of the twenty-first century, the use of animals to make "handbags and hamburgers" would seem not just crazy but archaic. The figure he provided for the "global herd" of land animals that supply us with meat and leather was sixty billion, not far out of step with the figures provided by the geographer Vaclav Smil in his work on meat and its relationship with the earth's animal biomass. Forgacs noted that by 2050 this could grow to a hundred billion land animals, supporting a population of some ten billion people, taking a substantial toll on our resources and our environment. Forgacs also cited a familiar figure, namely that raising livestock produces 18 percent of the greenhouse gases we release into the atmosphere.[12] In addition to its moral toll, to which Forgacs is not insensitive, our current industrial strategy for procuring meat and leather imperils "the environment, public health, and food security."

Fortunately, Forgacs went on to say, another path is possible, because "animal products are just collections of tissues." This may be the crucial and foundational view of cellular agriculture writ large. "What if," Forgacs asked, "instead of starting with a complex and sentient animal, we started with what the tissues are made of, the basic unit of life, the cell?" Some of these claims raise reasonable objections. The sentience of chickens, cows, and pigs is debatable; animal products may be "collections of tissues," but saying that they are "just" this belies their structural complexity. In the case of meat, "collections of tissues" leaves out the crucial dimension of time, the duration of months or years skeletal muscle takes to develop and mature, not to mention the dimension of space, or the environment (industrial or bucolic, cramped or relatively free) in which the whole animal lives. The phrase "collections of tissues" underwrites the notion that tissues can be duplicated without recreating the life histories of animals. But Forgacs wasn't authoring a paper or offering footnotes; he was representing the point of view of enterprise. It doesn't reduce the dignity of his talk to call it an excellent sales pitch.

As of late 2014 (and as of this writing) Modern Meadow's current goal is to make leather-like materials made from collagen protein, the same natural building block found in animal skin, and as strategy this is entirely correct. Compared to meat, skin is composed of fewer cell types. It is also structurally simpler, less reliant on three-dimensional complexities such as muscle fiber alignment or the layering of sheets of muscle. In terms of technical challenge, leather represents a lower hurdle for Modern Meadow's scientists to clear than meat does. And leather offers advantages as a product to bring to market. Things that we wear, Forgacs said, are "less polarizing for consumers and regulators" than things we chew and swallow. Modern Meadow will not need to guide its biomaterials through the same regulatory apparatus cultured meat would require. Forgacs called leather a "gateway material," opening the way to other kinds of biofabricated products, "without the animal sacrifice." Notably, the margins for leather are usually substantially higher than those for meat.

Forgacs moved on to an idea he shared with other proponents of cultured meat: a vision of cultured animal products being produced not through slaughter and butchery, but in sterile, sleek facilities that look a bit like breweries—a very widespread way of imagining cultured meat production. He closed on a more speculative and sweeping note: "Perhaps biofabrication is a natural evolution of manufacturing for mankind. It is environmentally responsible, efficient, and humane. It allows us to be creative. We can design new materials, new products, and new facilities." He continued: "We need to move past killing animals, as a resource, to something more civilized and evolved. Perhaps we are ready for something literally, and figuratively, more cultured." He had, by this time, held up strips of a leather-like material for the audience to see, early triumphs of laboratory bench work. One was black and opaque. Another, just a few sheets of cells thick, let the light through like stained glass. This is the precursor to what Modern Meadow will, in later years, present as its first brand of biofabricated material.

Forgacs was offering a point of view, but also a species of promise. Such is the nature of many talks of this kind, in which an entrepreneur, promoting a novel technology, tells stories about that technology's future benefits. It is sometimes unclear which of two common kinds of promise is on offer: "promise" in the sense of a new technology being "promising," or in the sense that someone "promises" to bring future benefits by means of technology? This lack of clarity has its uses. It can make obligations fall away when a promised future doesn't materialize on schedule. In other words, something about the ambiguity of promising is useful for managing the risks inherent in speculative research programs such as Modern Meadow's.

Presentations like Forgacs's TED Global talk are part of a genre important for understanding the conversation surrounding cultured meat. As Forgacs spoke in Brooklyn, my mind wandered, not only over that TED Global talk, which I've viewed on the Internet (along with over a million other people, judging from the website's view counter), but also over the commentaries that have been written about TED and the "big ideas" conference circuit of which TED is a part. In addition to TED and related conferences—including the local TEDx affiliate conferences (seemingly countless), TED Global, and TEDMED—a brief list would include the Aspen Ideas Festival, South by Southwest (SXSW), and others, all of which share certain common features.

The journalist Nathan Heller, writing in 2012, called TED "a showroom for the intellectual style of the digital age."[13] A luxury car showroom is not a bad comparison, for while many of the most popular talks in its history were not given by engineers or scientists, the conference has nevertheless been associated with technology, and with mobilizing energy and wealth to confront grandly scaled (and in fact global) challenges, often having to do with our natural environment, with human health, or with education, either in the developed or the developing world. The bigness referenced by the term "big ideas" indicates something decidedly non-ideational, namely the potential worldly impact of a particular novel technology or social practice.

TED and its talks have also attracted criticism, often targeting the conference's transformation of intellectual content into bite-sized messages that demand little cognitive mastication. Benjamin Bratton, a professor of visual arts at the University of California, San Diego, used a TEDx talk as a platform to criticize TED: "TED of course stands for Technology, Entertainment, Design. . . . I think TED actually stands for: middlebrow megachurch infotainment." This may have been intentionally mean, but it was only a more strongly worded version of a fear shared by some scholars and scientists, namely that TED, and other very influential "big ideas" conferences, prepare publics to value our work not when it is difficult or ambiguous, but only when it inspires them while promising to "solve for X." A kinder reading offered by Heller is that the big ideas talk has become a contemporary "sentimental" mode, consumed not only for informational or business reasons but also for loosely moral reasons, by viewers seeking inspiration. Some of these talks even play in the emotional register of Holstee's signage, as injunctions for mortals to take action in the world and pursue their particular passions, in response to the epiphany of our ultimate mortality. In a 2009 TED talk, Catherine Mohr, a designer of robotic surgical equipment, explains that life-threatening illnesses don't care "how many books you've written or how many companies you've started, that Nobel Prize you have yet to win, how much time you've planned to spend with your children." Her devices are intended to help heal people so that they can "go out and save the world."[14] One has to look past the clichés: the implication is that a novel technology temporarily corrects the failing body, providing at least a little more biographical potential for the person saved, a little more expression of their will in the world.

Even American National Public Radio has a program that airs TED talks, and the ubiquity of these "ideas worth sharing" (as the slogan runs) irritates many scholars who deal with less easily shared ideas for a living. For intellectuals to complain about intellectual shallowness at TED may seem like a minor turf war among elites. Yet the original founder of the TED conference, Richard Wurman, described with great economy an arguably deeper problem within his erstwhile endeavor: "They're up there selling 'doing good.'"[15] This point must be made with some care. Many TED speakers, or presenters at the Aspen Ideas Festival or at Google's Solve for X, are sincere enough. Wurman's "selling 'doing good'" describes a problem more complex than sheer disingenuousness: an optics of ultimate ends can blur the uniqueness and challenge of specific human endeavors. The roboticist references saving the world; a video game designer claims that the psychological incentives of game players can somehow be harnessed to secure social goods. A wide variety of experts, from linguists to geneticists to physicists to video game designers, occupy the same physical stage and end up crammed into the same rubric for salience, which turns out ("it turns out" is a phrase often heard in TED talks) to be a matter of impact, and poems and politics and protons are expected to hit the world in more or less the same way. A carefully lit stage can banish differences, whether they be the category differences that divide phonetics from photons or the differences of scale that divide tissue culture experiments from fast-food hamburger production.

The TED talk's status as a sentimental form may seem utterly incidental, but it matters for understanding the conversation around cultured meat. TED and TED-like talks, viewed over the Internet, have become one of the primary means through which audiences come to know about the effort to fashion artificial flesh. Not only that, but there is a particular affinity between the moral projects with which cultured meat is associated and the underlying premise on display in the "big ideas" circuit. Cultured meat is technological, innovative, a potential game changer—all the buzz phrases one hears outside conference auditoriums and in press releases. It is also designed to improve the world by moving through markets and appetites and digestive tracts. A consumable, its form resonates with the way ideas are rendered consumable on the conference stage. Later in the evening, an audience member will ask Forgacs if Modern Meadow is "just a pivot" for him and his father, a way for them to move from one business opportunity to another. This is unkind to Gabor and Andras Forgacs, but it is always worth asking such questions when someone proposes to create positive social change through the mechanisms of the market economy.

Up on stage with Andras Forgacs, I fumbled a few words about the social-scientific study of emerging technologies, and about the way foodways change over time. I was more eager to ask Forgacs to clarify a few things about his work, and I began with a question somewhat athwart his message. "So, if Mark Post wants to make a burger," I said, "it seems like Modern Meadow is less interested in copying existing forms of meat." Forgacs agreed. Modern Meadow had grown a number of what Forgacs calls "steak chips," and several of my contacts in the cultured meat movement have eaten them. They are still expensive to produce, though considerably less so than Post's burger, and Forgacs didn't bring any samples with him tonight. I went further: "But if the goal of the burger is ultimately to undermine the animal-grown version of meat by offering an exact replacement, what happens when we offer people forms of lab-grown meat that don't copy nature? Can they achieve that same end?" Forgacs was prepared with an answer, namely that what I called "mimesis" was unnecessary. If consumers choose a novel form of meat, their consumption of older forms of meat will inevitably drop. Alternative forms of meat were very much on my mind. Shortly before traveling to Brooklyn, I had received a copy of an art project called the *In Vitro Meat Cookbook,* produced by the Dutch design studio Next Nature Network, which examines the way the pursuit of cultured meat could lead to novelty rather than mimesis. Next Nature and its authors and artists collectively imagine futures populated by everything from "meat paint" (think fingerpaint) to in vitro oysters, grown in tiny shells like bioreactors, each perhaps permeated with an additive, oceanic *terroir* (the idea that the place where a plant or animal grows will register in the flavor of foods or beverages made from that plant or animal), via water from the North Atlantic or from the coastal waters off Japan *("mer-oir"?).*

My next question was a deliberate provocation, directed at the room. I asked if cultured meat reflects a cynical attitude regarding human nature, because it seems to place more confidence in tissue engineering than in our capacity to rein in our appetite for meat. "It's an interesting question, but a very academic question" was Forgacs's response. He went on: "At the end of the day, it would be wonderful if everyone in the world became a vegetarian or a vegan, but it's unrealistic." Acknowledging that many people (in the wealthy, developed world; in places like upmarket Brooklyn) try to eat less meat, partly for their own health, Forgacs nevertheless expressed reasonable doubts about this trend resolving the problems produced by animal agriculture. Any gains in North America and Western Europe, he said, would be more than offset by increases in animal product consumption in the rest of the world. He presented a version of the argument that has become the standard explanation for increasing meat consumption in the developing world: "As you become middle class, in emerging markets, the first thing you spend money on is better-quality food, more nutritious food, and that typically has been in the form of meat." At the moderator's prodding, he wrapped up: "We want to make products that allow people to make choices, and to make choices that can, in the aggregate, have a better effect."

The next panel featured representatives from organizations that promote local and environmentally sustainable agriculture, meat production in particular. Many of their messages harmonized with Holstee's "Food Rules," and its invocation of locality. Some of them also voiced another sentiment common among neo-agrarian food activists, namely that nature has its own kind of "intelligence." By contrast, they said, science's search for a silver-bullet solution to our food problems is entirely wrongheaded. One speaker suggested that the further we get from what we consume, the worse. Cultured meat seemed to her to keep animals and humans at a pretty distant remove. Forgacs responded diplomatically, stressing that Modern Meadow doesn't seek to be part of the industrial food system, but rather to produce products in an artisanal fashion (he noted that "one of our scientists is French"). Not only that, he and his opposite numbers in this debate shared the value of defending the environment, though they never dug into the question of just what constitutes environmentalism. At one point, one of the neo-agrarians got confused, seeming to forget that Forgacs and Post are different people, and talked about "Andras's burger." But the common thread, in the second panel, was that cultured meat is not the solution to our environmental, food security, or animal welfare problems. Better solutions, we were told, lie in those small-scale projects in ecologically friendly, and comparatively humane, animal agriculture that are beloved to the neo-agrarians. The problem of scale, and the challenge of feeding a hungry planet, seemed to have dropped beneath notice. A panelist whose primary work is to advocate on behalf of small farms complained that cultured meat does little or nothing to empower farmers. While I encountered little or no anti-agrarian sentiment in the cultured meat movement, it is true that most cultured meat research imagines food production from the standpoint of urban consumers rather than rural producers. And no one loves a narrative in which they, their skills, and the world they know all drift toward dust, while everyone else moves on to a better future.

A cook named Mike asked Forgacs, "Who wouldn't you take money from?" Forgacs, Mike implied, must be subject to the temptations of the money in big agribusiness. Forgacs insisted that Modern Meadow's investors are, for the most part, speculative venture capitalists rather than representatives of Big Ag. In fact, and though Forgacs doesn't put it in so many words, Modern Meadow is better understood as a small technology company than as an animal products company. Modern Meadow doesn't want to be purchased by Smithfield, Tyson, or some other very large meat concern. Forgacs's father Gabor, sitting in the audience, has already tried to clarify that Modern Meadow is not on the side of big agribusiness against small farmers, nor are they trying to kill conventional farming. Gabor went on to opine that any technology that can be developed will be developed, and in both good and bad forms. It's likely that both artisanal and industrial versions of lab-grown meat will emerge.

Mike's question helps us get a sense of the exact meaning of the food/science distinction at play here: it is not really about science per se, but about large-scale, industrial, capitalist agribusiness. Anyone in the room could accept that the history of agriculture is filled with objects we would have to accept as "technological," however grudgingly, from plows to carefully bred (or genetically modified) seeds. Furthermore, any of us could see the difference between the technological character of plant breeding and the technological character of genetically engineering plants. If pressed, all of us might furthermore agree that any food produced, bought, and sold within the regulatory regime of the U.S. government, subject to its agencies' standards for healthfulness and safety, is loosely "scientific" in the sense that scientists have helped set the standards to which our food conforms. But none of these agreements would get the neo-agrarians in the room to agree that cultured meat might be a good idea. Our arguments really have little to do with grounding naturalness by means of definitions, but instead have to do with industry, scale, and commerce. Cultured meat represents a very different kind of food production technology than any previously in use, one that severs ties with anything resembling a moral ecology of production.

I muse over all this as I watch the last panel from the beer table. The chefs have taken the stage, and the air is full of little sound bites pitched for volume, everyone layering their contributions over each other. "We eat too much meat, anyway." "Large companies can be ethical, too." "If cultured meat materializes, it'll be at places like McDonald's." "Why aren't the venture capitalists putting their money into restoring farmland?" "Because there's no money in that!" Forgacs starts to suggest that there's a direct parallel between making cultured food products and traditional types of fermentation, as in many fish sauces, and a food journalist in the audience calls foul, saying that while you could argue that a jar of Roman *garum* or Vietnamese *nước mắm* is a bioreactor containing a bacterial culture, such fermentation is a biological process of a different type than the cell culture techniques used in lab-grown meat. Forgacs has absorbed criticism from all sides, but he keeps a relatively cool head. Most of his contributions, at this point, are efforts to de-demonize Modern Meadow, against a crowd that insists on the demonic character of food biotechnology.

One of the neo-agrarian speakers describes agriculture as a form of posterity. We die, but farming and foodways keep going, making it crucial that we practice them and pass them on in the right way. The panelists begin to ask one another if they have children. The neo-agrarians—when I get grumpy I slip into the bad habit of calling them the Luddites—sing the virtues of an agricultural life lived in harmony with nature, down to the turning of the generations. Parenthood, one of them says, really wakes you up to what matters in life. I glance from the panelists to Holstee's "Life is short" signage and back again, part of my brain tuned in to a term of Sigmund Freud's, *Überdeterminierung* (overdetermination), whose technical meaning (in Freud's *Interpretation of Dreams*) is that certain dream images have multiple sources in the waking life of a patient, making their interpretation complex. In its most common colloquial use, however, "overdetermination" means something else. An overdetermined meaning seems so obvious that it obscures more subtle, and perhaps more important, meanings. I can't help but wonder if Holstee's signage is getting to us. Apparently, and perhaps not just this evening, arguments about whether or not certain foods are natural are moral arguments rather than scientific ones. This holds important implications for anyone looking for a common and un-obscured language in which to speak about the future of food. It shows our terms to be more prescriptive than descriptive, keyed less to the latest in scientific findings than to our values. Effectively "parasitic" on the modern industrial food system, as Laudan points out, culinary Luddism projects the ghost of a moral ecology and judges its host as inferior to that projected image, and all the while industrial production dwarfs its artisanal counterparts.[16] The postindustrial location of Gowanus offers a perfect vantage point for considering this phenomenon: artisanal food producers and sellers working in the shadow of something with which they could never compete. All of Gowanus is a tribute, if from some angles an ugly tribute, to an industrialization that radically raised standards of living, though not in ways that the land and water could support forever.

There is no argument, tonight in Gowanus, about the need to reform industrial agriculture as it stands. Our crowd eventually disperses in the soft rain. We are still divided over the question of whether reform should tap the resources of an imagined past or those of an imagined future. In a few short years, Modern Meadow will change its focus, ceasing to pursue meat and leather through tissue culture techniques, and pursuing instead the fashion and design potential of biofabricated materials. In 2017 the Museum of Modern Art in New York will acquire a t-shirt commissioned from Modern Meadow. Suzanne Lee, Modern Meadow's chief creative officer, will describe the shirt as an answer to a what-if question: How might we reimagine garments if they began with materials in a liquid form? Modern Meadow's goal will still be to create animal-product analogues from non-animal-derived surrogates, in this case by finding new ways to create collagen, the extracellular material that often functions like glue in animal bodies. If liquid bioleather could be produced at scale, ultimately creating an alternative to animal-derived leather, it certainly would represent a materials revolution, and perhaps a less controversial one than cultured meat: the naturalness of our clothing is a less contentious issue than the naturalness of our food.

ELEVEN

Copy

I think back to the plastinated disk of a burger in the Boerhaave Museum, pale twin to the one Mark Post used in his demonstration. Like its twin, it imitates a conventional patty. But I should be careful when speaking of imitations or of copies. I don't wish to imply fakeness, as if the actions of cellular metabolism, division, and combination to form functional tissue are any less real in a bioreactor than they are in a cow. It all depends on what you mean by "real" meat. Is realness in what feeds you? In chromosomal closeness? In grass-fed *terroir*? In some other, unnamed principle of somatic identity? But the in vitro hamburger is definitely a copy in one sense. It is a copy of a familiar form of meat, molded from a new ingredient with slightly different properties than—the word is tricky—the "original."

Mimesis (from the Greek *mimos*, meaning "imitator") has emerged as a keystone of making and marketing cultured meat, just as important as scaling up. In a 2017 promotional film for cultured chicken, rather different from the one Post used in his 2013 demonstration, the camera returns repeatedly to an artist drawing a chicken's feather. The hand holds a pencil, and the graphite marks are beautiful and clean on the page. This strikes me as a surprising choice, reminding the viewer of the difference between the seeming perfection of a mechanical or electronic copy and the imperfections of a hand-drawn rendering. Perhaps we are expected to think of artisanal food production, or perhaps the drawing is a fig leaf hiding a factory. It's tempting to cite Emily Dickinson's "Hope is the thing with feathers," but it seems certain that cell-cultured chicken would be grown sans plumage.

Post's expectations for cultured meat are premised on mimesis. Only when a cultured beef burger matches a conventional one in its bite and chew and price tag, he suggests, will consumers begin to choose the cultured one.

While there are examples of products marketed as ethical (such as fair trade coffee or free-range chicken) being priced a bit beyond their conventional counterparts, and selling, Post doesn't want to lean too heavily on consumer altruism. He reasons that customers will only make the ethical choice when it means keeping the diet they already enjoy.

In his well-known essay of 1936, "The Work of Art in the Age of Mechanical Reproduction," the German Jewish literary critic Walter Benjamin was preoccupied by the mass reproduction of artworks during the first decades of the twentieth century.[1] The "auratic" quality possessed by unique artworks, formerly experienced at a ritual distance, such as paintings seen in a museum, was suddenly diminished, as viewers examined those works on postcards instead. Spared a visit to the museum, they could touch these images of artworks, hold them up to the light, or stack them on a table. But copying was not without loss. The cultural historian Hillel Schwartz glosses Benjamin: "What withers in the age of mechanical reproduction is not the aura, the Happen-Stance, of works of art but the assurance of our own liveliness." The problem is not that anything has been taken away from the original, which in any case we can still visit, whether it is a work of art in a museum or a famous and influential building, the kind studied by trainee architects. Rather, the existence of copies makes us anxious about liveliness and originality themselves. Do we possess them? Do they exist at all? It is not the physical artworks themselves that undergo sublimation. We lose our certainty that we will know the difference between a presently solid object and the future air it will become.

An obvious objection is that hamburgers and artworks of the kind Benjamin describes are different kinds of things. One is food; the other is, at least for Benjamin, connected to our capacity for experience itself, and to the question of whether experiences (as opposed to natural objects) might be duplicated.[2] And hamburgers already exist in massive multiplicity, effectively copies of each other. Is aura even a relevant concept in such a case? But consider what Benjamin says about the aura in another essay, on the poet Charles Baudelaire: to invest an object with aura "means to invest it with the ability to look at us in return."[3] It is useful to not be too literal here. To imagine ourselves being examined by our art (or our food) is really to envision another angle on a moment of aesthetic appreciation, and to thereby thicken the experience of art. If copying causes aura to decay and thereby puts into question the "liveliness" of the moment of art appreciation or eating, then we have a clue as to what Benjamin might have thought of copied food. It is easy enough to understand the Slow Food movement, founded in 1986 by the

Italian farmer Carlo Petrini, in Benjaminian terms, as an effort to imbue food with "aura" once more, and thereby to revive the cult of food. In 1986, when Rome's first McDonald's opened its doors, Petrini, who speaks of a "right to pleasure," joined the protest against the fast-food chain, handing out plates of penne pasta to the protesters. One bowl of pasta, Slow Food proponents liked to say, was better than a hundred hamburgers, a phrase that neatly communicates Italian locality against the hamburger's multiplicity and its global rootlessness (that the hamburger is simultaneously global *and* American is one of this food's paradoxes). Those who cook in accordance with Slow Food dictates tend to favor the hand, and the roughening uniqueness it affords, over the machine.[4] Rachel Laudan offers an important rejoinder to Petrini: by the time he came around with his penne, she writes, "fast food joints" were an old tradition in Rome, "reaching back to the days of the Caesars." These were street stalls offering cheap, fast, and deep-fried foods, including the famous Roman doughnuts. Deep frying was a difficult business to execute in one's home, better left to professionals.[5]

Restoring a hypothetical aura, theoretically lost to food's industrialization, is not a major preoccupation of cultured meat practitioners. To ask for an "authentic" hamburger makes little sense in conversations about copying and scaling, nor are the forms of meat that cultured meat workers most want to replace (the fast-food hamburger, paradigmatically) usually rated highly for their "authenticity." Questions about mimesis are both technical, as in how best to produce the right kinds of muscle tissues and combine them with the right fats and other elements in order to approximate the taste of steak or chicken; and strategic, as in which types of meat are most important if we wish to win consumers over. Post's decision to make a burger was strategic in this sense—a sausage might not have captured international attention in the same way. There is every reason to think that Post is correct about mimesis, as he was about burgers.

But despite its importance for cultured meat's future, mimesis remains an unsettled question, ringed around by distraction and challenge. The challenge is technical, as indicated earlier. To reproduce the animal muscle and fat that we call meat has proven far more difficult than early claims suggested it would be. The distraction (I use the term in affirmation, not condemnation) is our impulse toward invention and play. Assuming we can turn cultures of muscle and fat and other cells into raw material for the creation of food, why remain attached to conventional forms? Why not learn to cook bricks of chicken breast, thin sheets of bone marrow, perfect spheres of pork loin, pyramids of trout? Certainly, one might object that vegetarian meat surrogates have been on the market in many parts of the world for many years, and they adhere closely to conventional meat forms, especially the hamburger and the sausage. No one asks why textured vegetable protein isn't widely available as a dodecahedron. The sheer sensational quality of culturing meat in the lab, however, combined with the technical difficulty of doing so perfectly, has inspired a range of alternatives, as natural flesh threatens—or, if you prefer, promises—to break with natural form. Someone fantasizes about growing translucent sheets of fish muscle to make see-through sushi; someone else grows turkey cells on scaffolding made of jackfruit to make a hybrid nugget. Despite the mimetic story line that drives the field forward, based on the idea of the biological equivalency of meat grown in vitro and meat grown in vivo, cultured meat nevertheless prompts questions about what it means to copy natural forms and what it means to break from them.[6]

There are many versions of copying, imitation, and doubling, including intentional ones (such as the production of dry ice volcanoes for grade school science fairs) and unintentional ones (like the birth of twins).[7] To begin with a philosophical version of imitation, in Plato's *Republic* mimesis gets a bad rap. The tenth book of the *Republic* includes an argument against representative art in general, grounded in the idea that the imitation of nature is ontologically inferior to new invention. In the *Republic,* the laudable alternative to mimesis is *methexis,* a way of relating to the Forms themselves (perhaps by building a table, which Plato offers as one example), whereas mimesis recalls the defects of phenomenal things as opposed to ideal Forms. A cosmologically elaborate rendition of this judgment on mimesis can be found in some versions of Gnostic cosmology, in which the creation of the world is attributed to a blind idiot demiurge named Ialdabaoth who effectively distorts a transmission from a realm of perfect Forms. But the version of copying that is closest to the bone, for cultured meat, is cellular copying, on which the entire technical enterprise depends. Tissue cultures grow through cellular copying, and one of the most common anthropomorphisms that crops up in cell culture experiments is that the cells "want" to divide and grow. But nothing divides natural growth processes from the ultimate form that tissue takes quite like cultured meat, since the ultimate form of the product need not match the animal body from which an initial biopsy of cells was taken. This very quality of plasticity, of growth that need not match the form it would naturally reach in vivo, may produce some of the uneasiness that surrounds cultured meat. Perhaps this plasticity is what inspired the rough jokes about creating meat from celebrities, or about culturing and eating our own flesh, all of which circulate predictably on the Internet. Malleability yields a creeping discomfort.

Is there a larger meaning when imitation tips into invention? In his essay of 1957, "Imitation of Nature: Toward a Prehistory of the Idea of the Creative Being," the intellectual historian Hans Blumenberg posed this question against an expansive swath of European history.[8] He argued that the modern attitude toward making new things—that is, both our enthusiasm for invention and the legitimation crisis we often experience in trying to live with what we have made—originates in a set of shifts in the character and meaning of human artifice. Technology, understood most broadly as "making," began as the imitation of natural processes but ultimately became the free play of invention disconnected from natural models. Blumenberg's story, which is admittedly a philosophical bird's-eye-view affair, begins in a kind of Aristotelian garden in which all technology is understood to either imitate or extend natural processes. In this garden, imitation is not only the reproduction of form or function. It also secures a sense of place within a natural order. The story ends in a modern age whose experience of its own technology is founded on a deep discomfort at having set ourselves up as creators. After growing drunk on power, we find ourselves hung over, but this hangover really derives from the cosmological dynamics of a prior world, one we exited in the process of becoming modern. All this demands explanation.

Blumenberg begins his essay with a description of a spoon maker, one of the main interlocutors of Nicolas of Cusa's 1450 *Three Dialogues.* Cusa's inventive spoon maker has not taken some natural form as his model for making spoons, but instead an ultimately human-made idea for a tool, held only in the human mind. Thus, the spoon maker emulates the spontaneous creative power of the divine. For Blumenberg, modernity is characterized by the human rebellion against imitating nature, and by the desire to set ourselves up as creators whose own creations are valid, but we pay for this by accepting a "groundless" existence. For mimesis was always first and foremost a matter of relation or connection to a world whose connectedness is not fully captured by the modern word "ecosystem," though the English prefix "eco" is derived from the Greek *oikos* (home or household). We are not at ease in our rebellion against mimesis. One of the philosophical culprits, prior to Cusa, had been the infiltration of Platonic ideas about the greater value of *methexis,* or the relationship between a particular object and the ultimate Form to which that object relates, over mimesis, into a regime of craft and creation that had been broadly Aristotelian[9] and either comfortable or unconcerned with the value of mimesis. In other words, Platonism effectively inserts a hierarchy of making that had not been there before, and this hierarchy was readily translated into the theological terms of rebellion, not against nature alone but against God, via imitation. Blumenberg argues, with some relevance to the impulses of technologists, that we moderns now experience *techne* (often translated as "craft") as a metaphysical event, and novelty as a metaphysical need. Yet a question mark hovers over our needs themselves, and over their propriety. *Homoiosis theoi,* the desire to be like God, is both compelling and hard to live with. The story of technology and civilizational imbalance has been told many times, and in many politically inflected ways. Blumenberg's addition to the story is that the problem lies not only in the effects of our creations on the world and on our bodies and minds, but also in the character of creation itself.

Cultured meat occupies an interesting middle ground in Blumenberg's understanding of the world. It appears to conform to the Aristotelian understanding of technology in some ways, for it "extends" the natural processes of the cell and of muscle development beyond what nature affords. At the same time, it provides such an obvious opportunity for flesh to take on new forms that only an intense attachment to the familiar ones could keep cultured meat within the paradigm of imitating nature. Many of the scientists who conduct experiments toward the production of cultured meat see the muscle cells they grow as identical to the muscle cells that develop in animal bodies. In the lab, though, they have to work hard to find the environmental conditions that promote cell growth and health, and the sheer effort involved gives them a keen awareness of our current distance from mimesis. They understand that at present, and perhaps permanently, cultured meat will only produce mimetic effects in its final form rather than in the laboratory. Despite the claims of many of its promoters, to make cultured meat is not "growing meat in the lab instead of in the animal," as if "growing" meant the same thing in each instance. The scientists and engineers with whom I spoke may not identify as Aristotelians, but they would not be surprised to hear that their biotechnology has introduced a crucial distinction between the Aristotelian categories of *natura naturans,* or nature as productive process, and *natura naturata,* or nature as a set of specific shapes.[10]

The bind of the mimetic, in the case of cultured meat, is that this entire philosophical dynamic must be pressed into the shape of a hamburger, when it so manifestly could overflow that mold. The imitation of nature, for

Aristotle, was a principle of relation, human hands making things that depended on their prehuman antecedents and called that dependency to mind. In early twenty-first-century tissue culture and tissue engineering, that dependence has become tenuous, and the presence of human will keenly felt. Blumenberg's argument in his 1957 essay was not that mimesis, and the sense of connection it might bring, would be better for us troubled moderns than invention, or that Aristotle is the key thinker out of whose work we should spin a history of *techne* in general or of technology in particular. His stakes were nothing less than our ability to view reality, whether organic or artifactual, grown or made, as legitimate, recognizing the human freedom called creation without getting caught up in a compulsive search for the next tool or toy. Blumenberg wished for a less tortured version of human artifice.

My interlocutors in cultured meat labs don't seem particularly tortured, at least not by the distinction between imitation and invention. The division between *natura naturans* and *natura naturata* in tissue culture work is mostly a door opened to new possibilities. The techniques that lead to the copy will (assuming the copying succeeds) inevitably lead beyond the copy, and eventually to the challenge of original creation, which Blumenberg might locate in some secularized equivalent of the human soul. The question is, when we create meat in the lab, what stories about us and our appetites will our creations tell?

TWELVE

Philosophers

"Guys, you're grandstanding at a panel that has Peter Singer on it!" Strange days when animal protection activists interrupt philosophical discussions about the suffering of animals, about which topic the philosopher Peter Singer is an expert of record. It is October 2014. I have joined a throng of people at the Manny Cantor Community Center on the Lower East Side of New York City for a roundtable on the future of protein whose question-and-answer session has become a three-ring circus. An organization called the Museum of Food and Drink (MOFAD) has organized a panel that includes Isha Datar of New Harvest and Patrick Martins, head of a heritage-breeds meat distribution company called Heritage Foods USA and the author, with Mike Edison, of a book called *The Carnivore Manifesto*.[1] The panel also includes Singer and another philosopher, Mark Budolfson. Singer's 1975 book *Animal Liberation* is often called the bible of the animal rights movement.[2] The moderator is Dave Arnold, a chef and professional innovator in food and drink. Arnold is well known in certain esoteric parts of the food world. A very short and merely representative list of his accomplishments includes founding MOFAD and, before that, a bar favored by cocktail connoisseurs, called Booker and Dax. Once a philosophy student and sculptor, Arnold more recently created a device he calls the "Searzall," effectively a set of wire meshes to place at the end of a kitchen blowtorch so that foods can be seared without tangling their flavors up in the aromas of burn gone wrong. It falls to Arnold to shout at the activists who have interrupted the event during the question-and-answer session.

Singer's *Animal Liberation* is thirty-nine years in print in 2014, and its rendition of utilitarianism stands behind the thinking of almost every philosophically engaged person with whom I spoke during my fieldwork in the cultured meat movement. The book remains a landmark argument in favor of freeing animals from our systems of food production and medical experimentation, as well as from experiments benefiting the cosmetics industry and sundry others. It begins not with love for animals but with a philosophical conviction that motivates animal defense. "I am a vegetarian," Singer has written, "because I am a Utilitarian."[3] So why have the activists chosen this event to stage a protest? "Peter, we appreciate you, we love you, but no one's talking about letting the animals live," says their spokeswoman. Her comment inadvertently limns the difference between Singer's approach to animal suffering and that of the activists. "Letting live" seems to mean allowing the animals to simply pursue their own individual lives, though the slogan includes no account of what those lives are like, save that they are as dignified as human lives. By contrast, Singer's is not a philosophy of the inherent value of the lives of animals. Human or animal life, for Singer, is always contoured by experiences of happiness or suffering. These, unlike notions like inherent worth, are the conditions a utilitarian can measure with some hope of improving the world. Philosophers sensitive to human experience are loath to condemn the emotional responses of sympathy and empathy, but Singer's approach demands that we begin not with the objects of our feelings but with arguments.

The activists, members of an organization called Collectively Free, carry signs that bear the images of animals humans eat: cow, pig, lobster. The signs say "I want to live," and the activists' chant is a call-and-response: "They want to live, just like you want to live, let them live."[4] Collectively Free's decision to disrupt this particular panel remains mysterious, for the speakers are scarcely Moloch-mouthed carnivores advocating the industrial-scale raising, slaughter, and mastication of animals. Patrick Martins does defend a version of carnivory, but this is the much-moderated kind of carnivory promoted by Slow Food USA, an organization Martins founded as part of the international Slow Food movement. Slow Food's complex program might be summarized as the rolling back of industrial agriculture and fast meals, and the elevation of the values of heritage and community. Martins is against the industrial-scale version of animal agriculture, which in his eyes is ringed around with the externalities of environmental waste and needless cruelty. He extols the virtuous meat-eating habits of Italian peasants (before Italy's industrialization) who, he tells us, ate perhaps an ounce of ground meat in a sauce, and that infrequently. Thus doled out, Martins says, a turkey might feed a village. Such small-scale consumption could be supplied by meat production on small farms, a more sustainable option from an environmental perspective. Very few meat promoters will tell you to eat less meat, but

Martins will. He will also gladly tell you that to ask poorer people to eat cheaper meat than rich people reflects a shameful class prejudice. He thinks that cheap meat must go, and that everyone should aspire to eat a relatively small amount of better-quality meat.

Datar is also a reformer. Her goal as head of New Harvest is to promote animal-product replacements created by tissue culture, and her contribution to this evening's talk is a vision of food production without animal suffering of any kind. As she describes New Harvest's work, I note her moderate tone. She presents cultured meat as a technological possibility well worth our investigation and investments of time and money, but not as a certainty or a panacea for what ails the food system. It just happens that at this particular event, the promise of cellular agriculture is drowned out by preexisting debates about the ethics of conventional, that is to say, in vivo meat production. Peter Singer is a reformer too, a philosopher who spends much of his time promoting utilitarian or, more broadly, "consequentialist" moral philosophy for the general public. Budolfson works within the same consequentialist philosophical framework but has reservations about Singer's arguments, especially where Singer promotes consumer choice as a lever of social change. Budolfson praises instead the use of government regulation to reduce harm to animals and the environment. One of his contributions to the discussion is the term "harm footprint," a twist on the idea of a carbon footprint that is infused not only with the desire to measure suffering, but to get us to take our moral failings as seriously as we take environmental damage, to see them as structural issues in our civilization.[5]

The panel does not respond gracefully to the activists' incursion, and I don't blame them. Collectively Free torments Martins, asking him how he would feel if he were about to be killed. He defends himself for a moment, perhaps under the impression that a reasoned debate is possible. He describes one porcine breed, the Red Wattle pig, which would not exist if breeders hadn't brought it back from near extinction to serve as a food animal. Then Martins, who could be forgiven for getting a little grumpy at all this rough handling, considers the slogan "I want to live" and says, "Well, I don't always want to live," implying not that he is suicidal but that from a human perspective, willing or desiring one's life is a complex journey punctuated by rough stops and jolting restarts. Singer suggests that it is unclear which animals, if any, know they have personal futures. It is also unclear whether our food animals are conscious of desiring that their existences extend over time. Survival instincts may just be instincts rather than a sign of a sense of existence. I recall that some animal protection activists have criticized Singer for not taking an extreme enough position on the sanctity of animal lives. Collectively Free doesn't just want near-agreement with their program, they want total agreement. Singer has not committed and will not commit himself to the defense of animal lives at all costs. Such a position would not reflect his views.

As a utilitarian, Singer is a proponent of a school of thought that emerged in the late eighteenth century in Britain. Classical utilitarianism (as it is sometimes called) has a legacy that is partly scholarly and partly political and social. Utilitarianism combines the following features. It is consequentialist (and is a subtype of consequentialism) insofar as it judges right and wrong by considering the outcome of our actions and not preoccupying itself with the nature of those actions themselves. It is a doctrine of ends, not means. It is universalist insofar as it claims to take into account every being's interests equally. It is welfarist in that it understands and measures people's well-being in terms of the satisfaction of their needs. And it is aggregative in that it considers everyone's interests added together, with the goal of maximizing happiness and minimizing suffering for the greatest number. Individuals count only as part of the whole. Each one counts for one, never for more than one.

If this account of utilitarianism's parts seems schematic, it is worth saying that many utilitarian accounts of the world can seem like line drawings or blueprints. As the philosopher Bernard Williams noted in an essay critical of utilitarianism, this approach "appeals to a frame of mind in which technical difficulty . . . is preferable to moral unclarity, no doubt because it is less alarming."[6] That is to say, for a utilitarian it is better to have a complicated job of balancing multiple interests than to be unsure what would count as a desirable outcome. Utilitarianism appeals to those who dislike moral ambiguity and to those who focus on outcomes, and this describes many actors in the world of cultured meat who eagerly anticipate an end to animal agriculture. It is a philosophy, we might say, for problem solvers both actual and would-be.

One of utilitarianism's chroniclers, the philosopher Bart Schultz, calls the early utilitarians "happiness philosophers." This name captures one very positive dimension of the project of William Godwin, Jeremy Bentham, and James and John Stuart Mill. In his essay of 1926, "The Harm That Good Men Do," the philosopher Bertrand Russell attributed many of the great social improvements of the British nineteenth century to utilitarianism's influence, from the Reform Act of 1832 (which made Parliament more reflective of middle-class as opposed to strictly aristocratic interests) to the Slavery Abolition Act of 1833, to the abolition of the Corn Laws in the late 1840s (which reduced food prices), to the introduction of compulsory education.[7] It is, then, curious that a moral philosophy seemingly so progressive in its powers, and so admirably on the side of the reduction of suffering and the maximization of happiness for all, might be viewed, decades after Russell, as a form of bureaucratic and restrictive reasoning antithetical to the freedoms and dignity of individuals. It is in the latter key that Michel Foucault wrote, "I hope historians of philosophy will forgive me for saying this, but I believe that Bentham is more important for our society than Kant or Hegel. All our societies should pay homage to him." This was sarcasm. Foucault was referring to Bentham's administrative invention, the "panopticon," an architectural design for prisons intended to ameliorate conditions in those institutions and aid in the correction of prisoners.[8] While Foucault's claim that "panopticism" was Bentham's real legacy does his target some disservice, it does get something right. From its inception, the utilitarian imagination was often (but not always) an administrative imagination. It presumes a fully disinterested perspective and assesses the respective happiness and suffering of different beings as if they were units to be administered. To accomplish this, it must establish some kind of equivalency between the disparate units it oversees. Singer deals with this problem by acknowledging that while human and animal needs are likely to be vastly different, both are subject to what utilitarians often term "satisfaction."

Satisfaction, in this sense, is an idea with analytic advantages. Though you and I undoubtedly have different needs, we both have a sense that our needs are being met a lot, a little, or not at all. However, it is worth noting that this principle of satisfaction can become a kind of smoothing function with regard to the different interests of individuals. If each one counts for one, and not more than one, the consequence is a strict refusal to attend preferentially to any particular person or any particular person's goal. "It is the greatest happiness of the greatest number that is the measure of right and wrong," Bentham wrote in 1776.[9] To Singer, this is the kind of "sound theory" from which we can build toward right action, rather than beginning with our moral intuitions (or, say, our sympathy for a particular creature) and then building theories that function to explain and justify those intuitions. A good utilitarian will notice his or her prejudices and remember to check them when it comes time to make a moral calculation. The problem with this "checking," as the philosopher Alasdair MacIntyre points out, is that it suggests that utilitarianism provides "no place for genuinely unconditional commitments," such as the ones parents make to their children.[10]

The name "utilitarian" came to Bentham in a dream in the summer of 1781, years after he had begun his important philosophical work. In this fateful dream, he founded a sect called the "Utilitarians," his written work having led directly to a community of shared belief and conviction. Bentham's vision expresses a more than passing interest in convincing others.[11] The paradox of classical British utilitarianism is that its "felicific calculus" contains impulses toward two things at once. First, it pushes toward social progress, subversion, and the dissolution of those forms of tradition that work against human betterment; and second, it shades off into a kind of bureaucratic reasoning that, especially from the standpoint of other versions of moral philosophy, reduces the complexity of moral determinations by rendering such concepts as "happiness" or "satisfaction" less fully dimensional than they might otherwise be. The utilitarian hopes to observe the tears and laughter of the world from the standpoint, as it were, of the universe.

Nor is utilitarianism particularly interested in a set of wider conversations about the philosophical consequences (to paraphrase the philosopher Christine Korsgaard) of living with nonhuman animals.[12] Korsgaard believes that animals present us with "a profound disturbance" in our thinking about the world, as if we "are unable to get them firmly into view, to see them for what they really are."[13] They are fruitful to think about because they are so hard to think with. Animals do not seem to organize their lives in the same terms in which we organize our own. A bear may recognize a berry as a good berry or a bad berry, but not through the same cognitive apparatus or within the same cultural milieu that causes me to recognize myself as a good son or a bad son. Both animals, bear and man, have norms, but they are different kinds of norms. A "strange extra dimension" distinguishes human life, a sense of life being a project we endeavor to work upon each day. The poet Paul Muldoon writes,

> Myself and Pangur, my white cat,
> have much the same calling, in that
> much as Pangur goes after mice
> I go hunting for the precise
>
> word. [. . .][14]

But to hunt for a word is not exactly like hunting for a mouse, and this is the tension upon which Muldoon's poem relies. The precise word is tied to other words and to a sense of expressive purpose that extends in time beyond the cat's killing a mouse and, perhaps, presenting it to a human companion. Just as poet and cat are alike but distinct, the quality of project-ness distinguishes my experience of wanting to live from a lobster's similar desire. This says nothing about the comparative validity of our respective wants, but it says much about their different characters.[15] However, it doesn't settle our moral problems concerning the treatment of animals to say that living with them raises important questions about human distinctiveness, unless we go so far as to brandish our distinctiveness as an argument for killing and eating animals. That is, unless we embrace anthropocentrism wholesale and decide that being human, and gratifying what appetites arise within us, licenses the cruel treatment of nonhumans.

At the core of Singer's *Animal Liberation* is the aforementioned problem of animal suffering. Singer's goal was always "expanding the moral circle" that surrounds those beings we deem worthy of moral consideration.[16] Singer did not claim that animal lives have the same value as human ones, but he also never staked a claim for the moral superiority of humans or the moral priority of their needs. Nor did Singer argue that either animal or human lives have a kind of intrinsic value, but rather that our (in the sense of "we beings") happiness or suffering is what makes us worthy of moral consideration. The validity of our aggregated present and potential future experiential states is what requires us to make difficult moral choices such as whether or not to raise animals for food.

Animal Liberation is less a book of philosophy than an activist's book that begins from a philosophical point of view. Its ratio of paragraphs of philosophical argument to paragraphs of detailed accounts of cruelty to animals in our food and medical research systems recalls the very dry martini: a vanishingly small amount of vermouth and a much larger quantity of gin. This is not philosophically invalidating (though some philosophers will object to my equation of philosophy with vermouth rather than gin). Arguments need not be long to be powerful. Singer wants to overcome a prejudice he terms "speciesism," or the belief that human concerns simply outrank those of other animals to the point where the suffering of other animals does not count. One of the most spectacular displays of speciesism in philosophy can be found, as Korsgaard points out, in Immanuel Kant's meditations on the origins of humanity, in which he considers the distinction between humans and other animals:

The fourth and last step which reason took, thereby raising Man completely above animal society, was his ... realization that he is the true *end of nature*. . . . When he first said to the sheep 'the pelt which you wear was given to you by nature not for your own use, but for mine' and took it from the sheep to wear it himself, he became aware of a prerogative which ... he enjoyed over all the animals; and he now no longer regarded them as fellow creatures, but as means and instruments to be used at will for the attainment of whatever ends he pleased.[17]

The idea that animals are mere "means and instruments," not worthy of moral consideration when weighed against "whatever ends" a human pleases, seems certain to offend Singer and his devotees. From the standpoint of the universe, the suffering of the beast is the same as the suffering of the human. Jeremy Bentham himself, in his *Introduction to the Principles of Morals and Legislation,* wrote of the possible distinctions between animals and humans from the perspective of the moral consideration of their suffering: "The question is not, Can they reason? Nor, Can they talk? But, Can they suffer?"[18] Singer chose the word "speciesism" to describe the denial of moral standing to animals, borrowing it from the psychologist and animal activist Richard Ryder, whom Singer met at Oxford. Ryder had coined the term in a pamphlet of the same name published in 1970.[19] The term's meaning is less philosophical than social-scientific—"speciesism" diagnoses a social prejudice akin to racism—but it can be understood as an application of utilitarianism's universalism to nonhumans. Again, to draw up a relation of equivalency between humans and animals in terms of sentience, defined as the capacity for suffering or happiness, is not to propose a more thoroughgoing moral equivalency between very different kinds of beings.

Singer was opposed to animal suffering in the industrial food system but not, perhaps curiously, to animal death itself. There was nothing in Singer's thought that compelled him to see life itself, or the lives of individual animals, as sacrosanct. Pain in life, he stressed, was actually more problematic than life extinguished. Singer noted that it is unclear whether many animals have a sense of their own ongoing lives and futures, which casts their deaths in different terms than the deaths of most humans. Nevertheless, he maintained that killing animals painlessly is an action safe from the charge of speciesism only if one is also willing to kill a human of a similar cognitive level to that animal. Singer maintained this view not to construe speciesism as morally wrong on its own terms, but because he understood speciesism to open the door to animal suffering.[20]

Singer's *Animal Liberation* has not been without its critics. Some simply deny the moral significance of animal suffering, but Singer's more interesting opponents have been philosophers who sympathize with the general project of defending animals but are skeptical of Singer's utilitarian approach.[21] One such critic is the legal scholar Gary Francione, who finds it puzzling that Singer (like Bentham) could oppose suffering but not see the death of animals as a form of harm demanding moral redress.[22] Francione reasons that Singer misunderstands sentience by foregrounding its seeming purpose, the registration of happiness or suffering. The actual role of sentience, says Francione, is to help an individual animal survive. To absent the death of animals from a calculation of the total harm done to them would be to evade both intellectual and moral responsibility. There is something to Francione's point, for creatures are not mere assemblages of immediate emotional and cognitive states.

Also prominent among Singer's deontological (that is, focused on the inherent rightness or wrongness of a deed, rather than on its outcome) critics has been the philosopher Tom Regan, author of *The Case For Animal Rights* (1983), in which he argues that the core wrong we inflict on animals is not suffering. All the pain of veal calves and boiled-alive lobsters only compounds the deeper wrong of the instrumental treatment of creatures that might be accorded their own rights. Dismissing contract theories of rights, Regan also rejected utilitarianism despite its attractively egalitarian qualities. "The equality we find in utilitarianism," he wrote, "is not the sort an advocate of animal or human rights should have in mind. Utilitarianism has no room for the equal moral rights of different individuals because it has no room for their equal inherent value or worth." This is a subtle point. From Regan's perspective, utilitarianism is laudably egalitarian but un-laudably disinterested in ideas like the worth of individuals, because it has decoupled individuals from their experiences (happiness, suffering) and decided that only the latter weigh anything. This converges with Francione's point that sentience is not merely the content of an animal's experiences.

After an early essay in which he argued that if certain humans (namely the mentally feeble and infants) have rights, then animals might also possess them, Regan went on to propose a rights theory based not on the idea of the contract but on the idea of inherent value, a quality that both humans and animals, all "subjects of a life," possess.[23] The implications of Regan's rights theory for animal agriculture are more stridently abolitionist (to use a historically freighted term common within the animal rights movement) than those of Singer's utilitarianism. Of course, a rights theory says little about how the defense of animals' inherent rights might play out in practice, or about what costs (real or expressed in terms of lost opportunity) humans might be willing to pay in order to see animal rights defended. The irony of the philosophical disputes over the moral treatment of animals is that most of the goals of the major disputants would be served by the same outcome, namely the end of animal agriculture itself. Many of the people around me at the Manny Cantor center, I learn by chatting with my neighbors, are abolitionist vegans who would like nothing better. Others are enthusiastic food lovers critical of industrial agriculture, who would be content with the mitigated animal suffering that small-scale animal agriculture could bring about, and who might be content to curtail their own carnivory accordingly, eating small doses of pig or cow from time to time and the occasional dollop of turkey.

The pressing question is whether there is a compelling moral-philosophical defense of the human practice of eating animals. Not a defense of eating animals under special circumstances in which human life is at risk (many utilitarians would condone these under emergency-airplane-crash-in-snow-blocked-mountain-range conditions; it would take an especially strict deontologist to disagree), but a defense of the practice of industrial-scale animal agriculture, or in other words a defense of everyday meat consumption, of cheap meat. As Singer points out, if such an argument were couched in utilitarian terms it would entail weighing human pleasure against animal suffering, and ascribing to each unit of human gratification a much greater weight than to the same unit of animal pain. This does, in fact, describe the world in which we now live, not because a moral judgment has been made but rather through the sheer accrual of historical episodes of animal husbandry, killing, butchery, eating, and (most significantly for a utilitarian) the massive upscaling of these practices in industrial modernity, a phenomenon everyone here at MOFAD's event, whatever their philosophical orientation, seems to regret.

Consider an effort to defend speciesism from the Singerian claim that it is morally corrupt. In 1978, philosopher Michael Allen Fox responded to Singer's argument as well as to an essay by Tom Regan, in the process folding their two rather distinctive approaches to animal liberation into one.[24] Acknowledging that animals may have interests, Fox disputed what he took to be the common claim made by Singer and Regan, namely that animal interests equaled those of humans. He furthermore denied that any rights-based claim could be made on an animal's behalf. Fox was not providing a philosophical defense of cruelty to animals, however, merely a defense of the idea that humans possess a kind of moral superiority over animals and that animals exist outside our moral circle. Fox suggested that identifying sentience with the capacity for suffering and enjoyment, and unifying humans and nonhuman animals under the umbrella category of the sentient, was problematic because it pursues a specific empirical ground for the possession of moral rights. Indeed, Fox argued, the attempt to ground specifically human moral rights in a set of universal and empirically observable qualities has often been frustrated by the sheer variation of human capacities. Moral rights have to come from somewhere else, Fox argued, and he claimed that even though cognitive capacity cannot ground our rights, our tendency toward autonomy (something secured by cognitive capacity) can ground them. Humans, because it is natural to our condition to have the capacity for moral autonomy, can be part of a moral community, and it is community membership that truly secures moral rights.

What is important here is not whether Fox scored philosophical points against Regan and Singer, which both authors denied.[25] What matters is that even in the midst of arguing for human moral distinctiveness, Fox did not think he had arrived at an argument that licenses cruelty to nonhuman animals. He did, however, think that the absence of a moral rights argument on behalf of animals licenses the human use of animals for food, so long as they are treated humanely. He acknowledged that there was nothing humane about factory farming. Fox's defense of what Singer termed "speciesism" was only a partial defense of carnivory, one that might cover Martins's method of meat production (providing it is as humane as Martins promises) but that would not serve to defend cheap meat.

As of this writing, I have encountered no philosophical argument that defends cheap meat satisfactorily, assuming we discount one reading of Genesis 9:3[26] according to which God gives all the animals to Noah and his sons (and thus to a restored postdiluvian humanity) as food. I can think of arguments for eating animals that are based variously on the legitimacy of cultural (including religious) tradition or on political arguments about the food sovereignty of peoples accustomed to raising and eating their own animals, but I cannot in good conscience say that these arguments seem to overcome the utilitarian and deontological objections to eating animals, and especially with eating animals who were born to be cheap meat. It is on these grounds that I view my own carnivory as a sign of my moral imperfection. Any moral-philosophical argument I might make on behalf of meat eating seems to me a fig leaf meant to preserve my gastronomic convenience.

Philosophy very rarely gets to meet the world of ideological contestation on its own preferred terms. "Have you read the documentation on crustacean nociception?" asks Dave Arnold, irate, his question shot at an activist holding a lobster sign. Has Collectively Free consulted the scientific literature dealing with the ways lobsters experience pain? An activist responds, "You have the right to sit around talking about killing them, so why don't they have the right to live?" Arnold tries to remind the activists that the panel is generally on their side, and then he gets understandably annoyed as they persist. "You're giving me a picture of freaking Babe the Pig here," he says, referring to the pastoral image of a pig in grass on one of the signs. "Please just ask a question," Arnold sighs, taking the microphone back from a colleague who may have grabbed it to keep Arnold from saying something he might regret. Singer tries to elevate the room's aggregate tone, saying that he appreciates the activists' message, and that there are genuine philosophical questions about the cognitive abilities needed to have a conscious desire to go on living. The pig or cow, Singer says, may well have those abilities; considering the lobster, he says he is more skeptical.[27] But they suffer, say the activists. Absolutely, says Singer, it's clear that they have that capacity. Martins takes the opportunity to surf on Singer's emphasis on suffering, saying that companies like his, which are committed to the happy lives and easy deaths of farm animals, are part of the solution to suffering—as, of course, cultured meat would be if it became a reality. "Everything wants to live," says Datar, "and if we could produce food without killing something, I don't see why we wouldn't do it." Singer has told the audience that he hopes meat won't have a future at all, and as the wave of applause breaks and subsides he amends this to say that he hopes it has no future except, perhaps, in the form of cultured meat. More applause breaks out.

But cultured meat, too, raises moral questions. Not questions about our moral regard for harvested cells, but questions about the implications cultured meat may hold for our moral regard for animals. Indeed, culturing meat may have more implications for our moral-philosophical views on animals than our moral-philosophical views on animals have for cultured meat. It is relatively easy to see how cultured meat would or would not suit different philosophical arguments for animal protection. The narrative about cultured meat Singer seems to support is one in which in vitro techniques simply allow us to eliminate much or all of our existing meat-production infrastructure, ending the suffering of millions of animals and the need to continuously breed more. A small seed population of each species might remain, numbering in the tens of thousands for the sake of preserving genetic diversity. The most salient philosophical objection to such an arrangement might be Tom Regan's, namely that the taking of biopsies of cells would be just one more instrumental use of animal bodies, and perhaps an infringement of animals' rights.

But assuming that cultured meat leads to the abolition of animal agriculture, it will change our sense of what these creatures, these nonhuman animals, are doing in the world. What are fully or partially liberated animals like? Despite our use of some of their cells, food animals in a world of cellular agriculture would likely be returned to their own devices. They would thus come into view rather differently than they do today, and Regan's phrase "subjects of a life" might seem to describe them better and better. If we ventured into the countryside, we might encounter animals more free to pursue the sort of personal potential that animals pursue, foraging and mating, raising young and growing old, socializing, and, in their own way, ruminating. The salient question is whether a more fulfilled pig or chicken would seem more deserving of its fulfillment than its pent-in, clipped, and unhappy counterparts. Would it seem like a creature with its own *telos,* much as we have our own? A creature that "wants to live" in ways that differ from the way we desire our own lives, but which we can recognize nevertheless? A creature whose individual experience, and not just the positive and negative feelings that ripple within its physical form, should matter as we debate how happy the world could be?

THIRTEEN

Maastricht

Why did cultured meat find its first stronghold in the Netherlands? Why were Dutch scientists among the first to embrace the idea and try to bring it into the laboratory? The easiest answer may be sheer coincidence: Willem van Eelen, a particularly insistent advocate on behalf of cultured meat, who late in his life organized researchers and won the government grant that led to Mark Post's burger project, happened to be a Dutchman. Post, the Dutch medical doctor, scientist, professor, and entrepreneur with whom I kept company whenever I could in 2014 and 2015, gives a different answer: the Dutch do not venerate their cuisine in the manner of other Europeans. Freed to regard their food as mere fuel, they are also free to think creatively about its design and creation. Much as they have thought creatively about their land, reclaiming about a fifth of it from the sea over the centuries.[1] You don't travel far in the Netherlands before realizing that the entire place has been effectively terraformed, every viable centimeter of soil rendered productive. Post's answer is impossible to test, but it harmonizes with the Dutch history of dikes and dams, and with their culture of meat eating that includes highly processed croquettes and sausages, in particular the *frikandellen,* whose carnal origins are industrially unclear but which are popular nevertheless. Post has eaten the same sandwich for lunch every day for years, a fact we should bear in mind when weighing his assessments of national cuisine.

I get to Post's lab in Maastricht by a roundabout route that begins in San Francisco, where I first meet him in 2014. He towers over me in a friendly fashion. I am tall by the standards of my family, but I stand just under the six-foot height that is about average for Dutch males, and Post is north of that.[2] We're both speaking as part of an evening of talks and conversation about cultured meat, held at the Dutch consulate in its offices near San

Francisco's Ferry Building, by the piers at the northeast corner of the city and close by the pylons of the Bay Bridge to Oakland. A year after his hamburger demonstration, Post is still the hero of the hour, and perhaps especially at a Dutch event. Post, in his fifties and vigorous, smiles frequently, jokes as often, and will not shy away from conversation on opera, which he loves. With a stunning view of San Francisco Bay visible through the consulate's windows, he tells me about ruined amphitheaters he has known on Mediterranean islands, and the views of the Italian coast from across the water in Messina. It is easy to think of Post as the very image of activity, a doctor-entrepreneur whose talents have finally reached a larger stage than the clinic and the laboratory. I learn that the Department of Expansion, the documentary company that made the promotional film used in Post's 2013 demonstration, made a second film, focused on Post rather than the burger, telling the backstory of the burger's creation. Though it was never completed or released, the filmmakers share clips with me, showing Post eating a meal with his family, doing sit-ups in the backyard, sailing on a boat. Since 2008, when he joined the Dutch cultured meat research project that van Eelen began, Post has become the face of an emerging industry, and his smile appears with a quick Internet search.

Our evening of talks is a collaboration between the Institute for the Future, in Palo Alto, and NOST (Netherlands Office for Science and Technology) Silicon Valley, an offshoot of the Dutch Ministry of Economic Affairs. It kicks off with dinner at Prospect, a restaurant whose sleekness stands out even in this posh corner of the city. The food at Prospect could be described as "California cuisine," and our large assembled group includes staff from the consulate, from NOST, and from the Institute for the Future, as well as the food writer Harold McGee. At the outset our discussion takes a somewhat surprising turn. We debate the virtues and drawbacks of what might be considered the ideological opposite of California cuisine. This is the meal-replacement beverage called Soylent. A young consular aide praises Soylent, which is relatively new to the market but increasingly well known in the Bay Area because of its popularity with, and association with, young male computer programmers who wish to spend less time cooking and sharing food with others, perhaps so they can work more, a weak form of transcending the body. California cuisine is more complicated. Ask three chefs to define California cuisine and explain its origins and you will get four slightly different answers, most of which will involve the local character of ingredients, the expenditure of effort, and the relationship between a given plate of food and a very specific network of regional farmers and gleaners. "California cuisine" is a high-value term, and the restaurants identified with it are comparatively expensive. Soylent, by contrast, comes as a dry powder or prepared slurry and features an anonymous ingredient list. California cuisine and Soylent both signify in elite registers, albeit ones with very different ideas about the links between food and land, and food and sociability, not to mention the proper uses of human time. One advertisement for Soylent suggests that the consumer hold a video game controller in one hand, a bottle of the product in the other.

To criticize Soylent while sharing dinner with well-dressed and sophisticated colleagues at a restaurant like Prospect is cheap sport. We may dismiss Soylent as encouraging a form of contemporary anomie, a nearly dystopian gastronomic loneliness. And yet the consular aide makes an intriguing point: the arc of gastronomic history has, through industrialization, bent toward spending less and less time preparing meals. Soylent would represent a plausible terminus ad quem for this trend, albeit one not all of us embrace because of its apparent molecularization of food, its implication that where wholeness was, there shall nutrients be. As Rachel Laudan argues, the modernization of agriculture and food processing has created a world in which cooks—who, given gendered divisions of labor, are mostly female—spend only a fraction of the time their grandmothers likely spent in the kitchen.[3] In any case, it seems silly to assume that either California cuisine or Soylent represents an option for human subsistence in general, rather than for certain rarified populations, some of them wishing for food to be richly imbued with symbolic value, others comfortable with it being purged of the same.

Soylent takes its name from "Soylent Green," the mysterious foodstuff featured in the 1973 film of the same name, which was based in turn on Harry Harrison's novel *Make Room! Make Room!*, a 1966 Malthusian fantasy about overpopulation.[4] Soylent Green "is people," as the movie's most famous line says. The green wafers are made of processed corpses. This raises questions about just what message the creator of the real-world product, Soylent, wished to send. His company has tapped into another dystopian fantasy by naming a coffee-flavored Soylent beverage "Coffiest," after the coffee surrogate drunk by the corporate wage slaves of Fredrik Pohl and Cyril M. Kornbluth's *The Space Merchants* (1952), a novel that features one of the most striking literary depictions of cultured meat. It is unclear how much more tongue the cheek can accommodate.

Post is heading to Los Angeles tomorrow for a presentation to the Roddenberry Foundation, and he and I agree it would be fitting if the foundation established by the son of Gene Roddenberry, the creator of *Star Trek,* supported laboratory-grown meat. Post mentions the "replicator" of *Star Trek: The Next Generation* (1987–1994). The replicator is a science fiction technology that creates food and drink almost out of thin air, through minute and deft molecular assembly. Characters often speak of "working on a recipe," which means reprogramming the computer to balance, say, nutmeg and ginger in warm milk. Whether or not there is anything Dutch about cultured meat, much of the money and inspiration for Post's work comes courtesy of California backers and Californian imaginations.

If Post is the entrepreneurial scientist who put cultured meat on a plate, Willem van Eelen (1923–2015) was its human catalyst. In the 1990s, after nursing the idea for decades, van Eelen, not a scientist himself, partnered with medical researchers to develop an in vitro meat process, filing patents in both the Netherlands and the United States but not achieving any great success in the laboratory. In the mid-2000s, van Eelen, increasingly impatient to see his vision realized, collaborated with Henk Haagsman, a professor of veterinary medicine and meat scientist at Utrecht University. Haagsman would become the principal investigator for a substantial research project, which connected Haagsman and his Utrecht colleague Bernard Roelen with Carlijn Bouten (Eindhoven University of Technology) and Klaas Hellingwerf (University of Amsterdam). The grant they received, given by the Dutch government agency SenterNovem, lasted from 2005 to 2009. It was in 2008 that Post, who was then based at Eindhoven University, joined the team, having taken up Carlijn Bouten's responsibilities. As he acknowledges, Post is thus an inheritor of Dutch cultured meat research rather than its initiator.

Post sometimes says that he got famous by accident. He happened to be the researcher available for an interview at the precise moment when a journalist was writing a story for Reuters and the Associated Press, and things rolled from there. Even if he did not seek out fame and the attention of image makers, Post obviously enjoys the stage, even the small one provided by the Dutch consulate in San Francisco, where our group spends the rest of the evening. The sweeping northerly view of San Francisco, the East Bay, and Marin is one of the best I've seen in my many years of living here, and I recall a brace of lines from Hans Blumenberg: "Summit meetings will retain their nimbus even if they often produce nothing. A form of superstition is connected to the highest authority: from such authority, it must be possible to make arrangements both against calamity and for salvation, precautions that no one else could imagine or even take responsibility for."[5] Highest authority?

Perhaps not, but we have height and a view, and a shared interest in preventing catastrophes at a global scale, although it is not entirely clear that we agree about just what the catastrophes are. While we all agree that climate change is a real and pressing threat, opinions about the significance of animal suffering and the future of food security vary. The forty-eight-story Transamerica building, a thin commercial ziggurat, is clearly visible, and so is Coit Tower, first opened in 1933, perched on its distant hill. Against this backdrop, as the sun sets, the Dutch consul makes sweeping opening remarks about the importance of a pyramid of "profit, people, and planet" that he hopes we will collectively create, and he invokes the Iroquois ideal of making future plans that keep in mind a full seven generations of our posterity. He then apologizes and departs, having earlier mentioned that his daughter is flying in from Amsterdam to attend the Burning Man festival in the Nevada desert, and that he must run to pick her up at the airport.

Post addresses some forty guests. At the heart of his presentation is the increasingly familiar puzzle of our relationship with meat. Humans seem to enjoy eating meat, but they have no obvious nutritional need for it, as evidenced by the millions of healthy vegetarians in the world. A desire only loosely linked to dietary requirements seems to have us in its mysterious grips. Post shares the view, widespread in cultured meat circles, that the mass adoption of vegetarianism is extremely unlikely, and that we must hope that technology can accomplish what behavioral change will not, for industrial-scale meat is contributing to our common doom. Post is not a professional salesman, but a scientist whose transparency regarding the technical challenges at hand, and the slow pace of progress toward clearing them, is laudable. Nevertheless, an audience member with little exposure to the topic could easily walk away with the impression that technological hurdles are much less significant than the eventual challenge of consumer acceptance. Post says that he hopes to find an economical, non-animal-based growth medium (to replace fetal bovine serum) within the year, news I receive with surprise, and he goes on to describe cows in a mechanizing way I have heard often: cows are an obsolete technology, inefficient converters of feed calories into edible muscle.

Post would like to replace those "obsolete" cows with bioreactors, each with a capacity of twenty-five thousand liters, each producing enough meat to feed forty thousand people. You could not swim effective laps in such a bioreactor if it were filled with water; twenty-five thousand liters is about the capacity of a small tanker truck. Post also introduces us to the more-efficient-than-normal cows from which he and his team source stem cells: the

Blanc-Bleu Belge (or Belgian Blue), a breed with a mutation advantageous for muscle production. Normally, mammals produce a protein called myostatin that inhibits muscle growth. The cells of the Blanc-Bleu Belge produce no myostatin, which means that the animals display double the muscle growth of other cows and are prized for their meat. It also means that Belgian Blue calves are already too large, at birth, to pass through their mother's birth canal and have to be delivered via cesarean section. Nevertheless, the muscle cells of the Belgians do not bulk up on their own in vitro. They require artificial stimulation by electrical, mechanical, or chemical means. Post details the painstaking process by which the muscle fibers for his burger were created, and I cannot help but wonder if I am watching an analogous process, in which this audience is itself stimulated to see in vitro methods as a legitimate source of meat. One way to do this is to describe cows as machines. This makes it easy to suggest that cell growth and muscle development in bioreactors can be more or less like the same processes in vivo.

Post is describing early polls of British and Dutch consumers indicating that large numbers of people might be willing to try laboratory-grown meat, when an audience member, a self-identified vegetarian, interrupts him. She disagrees with Post's earlier assertion that mass vegetarianism is unlikely (though neither of them offers an argument to back their claims). Trying to change the terms of the conversation, she asks, "Why don't you just tell people about the problems with the meat they already eat?" For the first time all evening, Post is temporarily without an answer, but he recovers quickly, responding that even though mass vegetarianism would be a reasonable response to the problems of industrial animal agriculture, we make our food choices on the basis of emotion, not reason. He mentions the Dutch *frikandellen,* noting that they are popular despite being widely acknowledged to be terrible on every level, from health to flavor to their industrial origins. Emotional response is the reason Post considers it important to practice "beef mimesis," as he puts it in passing, copying in vivo cow tissue whenever possible. Gastronomic inertia is powerful, and thus it is important to appeal to people's existing tastes rather than propose that they develop new ones. Post gets his round of applause, and we segue to other speakers, some of whom do hope that we develop new tastes—these are purveyors of insect flour, dairy-free vegan cheese, and whole crickets. Our evening concludes with a discussion of the likelihood that animal agriculture will have to change as global warming reduces available land, not to mention making life harder for the animals themselves, who will have to devote valuable and finite energy to keeping their body temperature as low as possible. The vegetarian's objection is the only one we have heard. No one questions Post's main premise, which is that the problem of the future of animal protein can be resolved through technology, and perhaps more easily than through any form of social change.

The river Maas (Meuse in French) bisects this city, which began as a Roman settlement when the Maas was crossed in Latin: Traiectum ad Mosam, or, much later and in Dutch, Maastricht.[6] I cross the Maas one morning on a jog over the oldest bridge in the Netherlands and find a flock of sheep sitting contentedly in the median of a major road. They are fenced in and their shepherd is nowhere in sight, nor is anyone else, it is so early in the morning. I am unused to meeting flocks of sheep in cities, and I stare happily at them for a moment, bouncing in place so that my legs don't get cold. I think of the sheep and cows I spotted on my train ride here from London, by way of Belgium, my car filled with people I took to be U.S. tourists, many of them speaking with Texan accents, all closer to seventy years old than to thirty. After pulling out from under St. Pancras station's graceful steel roof, we had passed fields golden with mustard, sheep and cows in pasture, the vestiges of a less densely compressed animal agriculture. One man turned to his companions and told a wistful story about growing up in Texas before the rise of the big feedlots. "That's agribusiness," he sighed, regretting the fall of a world in which he might see individual animals he recognized on a walk home from school. I grew up as a city kid and lack any such memories of my own. I've never been watched by so many sheep.[7] I worry that I look predatory.

In Maastricht I find a superficial lull. Post is occupied with the job of being a professor and administrator, as chair of the department of physiology, and busy with the daily life of his family. The technicians and assistants and volunteers connected with his cultured beef project are at work on their experiments but are not presently making meat. If I had expected to find a small factory turning out beef patties, I would now be deeply disappointed, but I have become used to the idea that I am in a story of incremental progress rather than dramatic advances. Meanwhile the workaday equilibrium of the Post Lab is often interrupted by media requests. As of early 2015, Post is still the first person journalists call when tasked with covering cultured meat. The sense of a lull is deeply misleading, however. Post is in conversations regarding the start-up he is cofounding, called Mosa Meats. He is in other conversations about creating leather, based on an inquiry from a major sneaker company. Interns and students move through his lab conducting experiments, some of them international arrivals excited by the possibility of tissue culturing meat. Post hasn't slowed his work so much as he's become focused, out of necessity, on the administrative and fund-raising and business sides of his enterprise. I'm catching him at a moment of transition between demonstrating that his techniques are sound and effective in the lab and demonstrating that they can move from the lab to the wider world.

Maastricht is a small, cosmopolitan city of some one hundred thousand citizens. The provincial capital of Limberg, it has been a crossroads of sorts between Belgium, Germany, and Holland for centuries. Maastrichtian natives built on the site prior to the Roman encampment, but the Romans seem to have been the first to mine chalk, as well as flint and limestone, from the nearby Sint Pietersberg (or Mount Saint Peter), bringing out enough chalk to build a sixteen-foot-high wall around what would later become Maastricht. There are traces of Roman ruins if you know where to look.[8] I'm told that the local dialect is closer to German than the Dutch I've heard in Amsterdam, and my ear corroborates this out on the street. The walk from my hotel to the University of Maastricht is long and takes me between many large, late-twentieth-century buildings that form tunnels for the cold spring wind to rush through. The university, established in 1976, is architecturally uncharming and functional and sits on the outskirts of town. When I reach the university, I try to look as sympathetically lost as possible and find my way through concrete and glass blocks, helped by the occasional student.

Vivian Schellings, the departmental administrator for physiology, leads me in to Post's office to wait for him. Spare, it nevertheless shows a few signs of Post's extracurricular interests: there is a poster advertising a production of Puccini's *Madam Butterfly,* on which a stylized Japanese woman poses holding a fan. There are also posters for museum and gallery shows of the painter Simon Chaye, and of Rembrandt, the latter from the Metropolitan Museum of Art in New York City. I recall that Post spent years working in the United States—six years at Beth Israel hospital in Boston, when he lived in the neighborhood of Beacon Hill, and a stint at Dartmouth-Hitchcock in Hanover, New Hampshire, as well. There are two desks, one with a computer and another that Post uses for meetings with students, a chair on each side. Post arrives, and our very first conversation is about privacy. I show him a form I countersign with interviewees, which stipulates the terms on which I'll quote statements made in formal interviews. He dismisses the form as unnecessary. He says that privacy is on its way out, something he will repeat several times during my visit. I admit that I'm puzzled, although I don't pursue it. Is Post referring to the sheer amount of data about us that Internet companies gather? His interest in transparency can't help but remind me of the Latin meaning of "in vitro": under glass.

We talk about the week to come, during which I will shadow Post whenever possible. Besides his teaching, advising, and laboratory schedule, there are several special events, including visits from two television news crews, one of which fails to turn up, perhaps to the relief of Post's often-interrupted lab team. Post will also give a talk to a group of medical students at a conference center in the nearby city of Nijmegen, not far from Eindhoven University, and serve as part of an expert panel on a local television program, where he regularly appears. Post laughs at how busy he is. In the Netherlands, he says, an academic like himself is basically a civil servant, obligated to do many things beyond his professorial duties. Post leads me on a tour through the laboratory, including the rooms in which his team crafted the cultured beef burger. I see the hoods under which Post and his technicians labored with T-flasks and growth media and pipettes, and the microscopes under which they examined cells forming muscle fibers. Conference posters on the walls of the hallways remind me that most of the research done in Post's lab is medical, and done under the auspices of the Cardiovascular Research Institute Maastricht.

There are very few signs of the work that was done on the hamburger—no celebratory banners or posters left over from the London event—but the door to a shared office, where I'll have a little desk space during my visit, bears a sign reading "I am a Meat Scientist" in English, and a t-shirt hanging on the wall of the office reads "kweekvlees," which is more or less Dutch for "cultured meat," and continues, in Dutch: "What is your ultimate meat? Ostrich fillet. Product information: prepared in the lab, not genetically engineered, no bovine encephalopathy." A touch of cultured meat's capacity to generate fantasy there—what animal would you eat, if you could culture the cells of any creature? There are few other traces of food culture, save for a few packages of instant ramen noodles, presumably not intended to serve as cell scaffolding. Folders of experimental protocols sit on the shelves next to instruction manuals and lab equipment catalogs. I learn that much of the lab's work involves heart failure in large animals such as pigs, whose cardiac tissue resembles that of humans. The lab does a great deal of work with animals, and Post says he is untroubled by this, so long as humans benefit from the research. I get to see one of the bioreactors that Post intends to scale up for industrial meat production. It looks like an Erlenmeyer flask with rotator blades in metal, inside, connected to tubes that circulate medium into and out of the bioreactor. I imagine what one would look like if it were scaled like a tank in a brewery.

Anon van Essen is one of the technicians who made Post's burger a reality. He has been full-time on the project since 2012. "I liked the idea," he says, in halting but proficient English, "to make hamburgers out of cells," in order to eat fewer cows and to "be more green." What he made "looks like a hamburger, and tastes a little bit like a hamburger," says Anon, not himself a vegetarian. Trained as a tissue engineer, he is now searching for a replacement for the fetal bovine serum used to grow the original burger. There are hundreds of commercially available candidates, many of them too expensive to use at an industrial scale. The trick is to find the ones that match the particular satellite cells to be grown, and then see if there's a way to reverse-engineer them to devise cheaper equivalents. Anon reminds me that the project's other current technical needs include adipogenesis, or finding a way to produce the right fat cells for the burger, and developing the proper scaffolding or beads for the cells to grow on, probably working with an alginate base. I ask if he can predict when a burger might be available at commercial scale and fast-food-competitive pricing, and he says fifteen or twenty years; the next day, I hear Post suggest three to five, fewer than he had predicted in 2013. I wonder to myself if spending a great deal of time at the laboratory bench, confronting technical challenges every day, makes targets seem further off, but I am mostly delighted that Post has made no effort to get his staff to parrot his predictions. Anon's press badge from the 2013 London event is right above his desk; he attended even though his son had just been born.

I have fragmented conversations in the hallways. Daniel, a longtime post-doctoral fellow in the Post Lab, speculates about how the Dutch got so tall, gesturing toward salmon-rich rivers from which everyone used to eat freely. Daniel is skeptical about the widespread contemporary desire for all food to be "natural," an unscientific label if ever there were one. Apropos of another kind of label, he mentions that the Netherlands is actually a prominent site for the production of Parma ham; pigs are raised here, then shipped south to Parma for a residency long enough to "finish" the pigs on *terroir*-granting local feed and to slaughter them on Italian soil. Michael, a volunteer technician who is plowing through Anon's growth medium data, thinks it's very likely that they'll find a substitute for fetal bovine serum, but not one that quite matches its effectiveness. He expects a vegan substitute might provide

80–90 percent the same rate of growth, something I have heard corroborated by researchers at other labs. A young researcher named Daan, a visiting student intern currently working on his master's degree in a combined program in biotechnology and business, seems intent on impressing me with his enthusiasm. His current task is to find the right ratio of microcarrier beads to cells, in order to get the cells to proliferate in the desired way. Daan is frustrated by European public attitudes toward the application of science in food, and he mentions that the application of genetic techniques could help with many aspects of cultured meat production, including growing fat cells.[9] Marco, another young intern who is doing exploratory research that might lead to the production of leather via cell culture, is looking at the role of fibroblasts in growing skin. He hit a common tissue-culture pitfall a few weeks back: a fungal infection made it necessary to destroy a whole fibroblast culture. This is a problem that Post has speculated he could avoid by creating perfectly sterile facilities, in which robot workers tend tissue cultures, eliminating the risk of such infections.

Two days later, I have settled into a routine of taking notes and observing experiments when Anon comes in to tell me that the German television crew from Deutsche Welle is here. In comes Andreas Hausman, the director, along with his cameraman and a younger assistant who will do the tiring work of holding the boom microphone during the shoot, which will be conducted in excellent English. The point, I gather, is to recreate the process by which Post's team created their burger. Anon, wearing a lab coat, sits at the lab bench and unwraps a small package of beef. The crew trains their camera on him as he takes a small fragment of it and places it in a dish along with some solution. This is the process of breaking up the tissue in order to cause the effectively dormant stem cells in the sample to become active. These are satellite skeletal muscle cells, the kind that Post's lab used to create their burger, and after they proliferate they will form the muscle strands that will become meat. Such skeletal muscle cells play an important role in vivo, which is to repair muscle when it becomes damaged. When the muscle sample from a very recently killed animal is cut, these still-living stem cells begin their work. Cultured meat begins with the mechanisms for muscular repair, a healing process becoming a potential production process.

Andreas asks Anon how long it takes to grow enough material to make a single burger. "About two or three months," Anon answers, using the techniques he's now demonstrating. This is still an artisanal process, which would have to become an industrial one to be viable. "Do you think this is the future?" Andreas asks, and Anon says, "I think so, yes." We take a break for a few still photographs. Anon shakes a test tube with a bit of meat in it, for the camera. How many cells are there in the burger, asks Andreas, and Anon answers that there are many strands of muscle tissue, each of which contains about 1.5 million cells, so there must be billions of cells in the burger, from an initial sample of only thirty cells.

By this point Post has entered, greeted Andreas, and donned a lab coat. In the next step, the sample will get spun in the centrifuge. "It's not spectacular," Post says, half apologetically, meaning it's not visually spectacular, but the centrifuge does have a name, Gentle Max, in honor of its slower rate of spin. Post makes a few general remarks for the microphone: cultured beef would, theoretically, require far fewer resources than conventional beef and produce more food value in exchange for those resources; it would produce fewer greenhouse gases than conventional cattle ranching; it would spare the lives of cows. Post breaks into a kind of mini-lecture, in which he covers the recent history of cultured meat research in the Netherlands, and Andreas asks him if he is as obsessed with cultured meat as van Eelen was. He gets a smile in response. "No, not as obsessed," says Post. He likes to present himself as levelheaded. He is straightforward about the fact that the tasters of his 2013 burger did not pronounce it the equivalent of a conventional burger. There is still a great deal of work to do.

A health twist I hadn't expected: Post acknowledges that certain vitamins, such as B12, have to be added to cultured meat, because unlike in vivo muscle, a tissue culture can't absorb B12 from surrounding tissue. But, he goes on, we also have an opportunity to enrich cultured meat, perhaps by inducing fat cells to produce omega 3 fatty acids, in order to lower cholesterol. There is relatively little discussion of the possibility of making meat healthier by modifying it in these ways, perhaps because to some audiences it may be too reminiscent of genetically modified foods; healthier meat, in other words, would also be less mimetic meat, and more potentially off-putting. With the camera rolling, Post says that consumers will absolutely buy cultured meat products. I keep backing up to avoid the boom operator as he, too, weaves backwards.

We move out of the lab and into the university halls for exclusively narrative reasons. Andreas wants footage of Post walking through the science library (which Post's lab work and teaching gives him, he admits, very little reason to visit; he hasn't been here in years) and arriving at the university by bicycle and locking up at a rack. This is a Dutch action sequence if ever there were one, I think to myself. It is important, Post says, to make sure that your bike lock is more expensive than your bike.

We get into the Deutsche Welle van and drive to a burger restaurant in downtown Maastricht, not far from the river. The wife of the chef is the daughter of one of Post's neighbors, and Post eats here often. The restaurant's interior, though, is less small-town than smart casual. We could be at a burger joint in San Francisco or Boston or Chicago, the kind of place where blue cheese, caramelized onions, and perhaps non-bovine meats like turkey or lamb are readily available. Andreas has Post order a beef burger, which is called a "Black Tiger." With the camera on him, Post eats, and says a little more about the timeline he anticipates leading up to a fast-food-competitive burger. It may take, he says, three to four years to produce a burger that could be served in a restaurant, and then another seven to eight years to reach a burger competitive with the major fast-food chains. As the Black Tiger drips sauce and melted cheese on his plate, Post observes that the bun is not quite up to its task of containment. Andreas then gets him to use the burger he's holding as a prop to illustrate his points: this burger would be better, Post dutifully says, if it took fewer resources and involved less pain and suffering for the animal killed.

Post reports a conversation he once had with the chef Ferran Adrià, of the now closed but very influential restaurant elBulli, about the possibility of Adrià being the one to cook his hamburger on camera in 2013. Adrià apparently considered it but concluded that his presence would not be good for the demonstration, and he offered an intriguing explanation. If Adrià touched the meat, it would transform the hamburger's culinary value, as if through a transitive property of his culinary reputation. It would distract from the idea that cultured meat can be a solid and dependable ingredient in its own right. I'm reminded of another prominent chef's response to the idea of cultured meat, at one of the many media events I attended during my research. Like Adrià, this chef is associated with "molecular gastronomy," a label not much loved by the chefs to whom it is applied, but generally indicating the use of sophisticated equipment, sometimes borrowed from laboratories, to transform ingredients beyond their familiar forms. Foie gras might be aerated into an offal cloud, or eggs Benedict might consist of breaded fried cubes that break into yolk on the fork. This chef stated he would not cook with cultured meat because of the ingredient's apparent lack of "integrity." I wondered at this insistence on integrity in a genre of cooking that often transforms things beyond resemblance to their origins. Was it prompted by a concern for a certain line between nature and culture, in which ingredients (produced over agricultural generations through plant and animal breeding) are taken from nature and become culture through the agency of the chef? Was the real motivation control—in other words, a belief that the chef should get to decide what is nature and what is culture?

We thank our hosts at the restaurant and then move again, this time to a spot not far from St. Servaasbrug, the footbridge that crosses the Maas, called the oldest in the Netherlands and rebuilt many times since Roman days. Post poses with the river at his back, and Andreas asks him to repeat his major points once again. I wonder how many similar news crews have come to Maastricht, and how many landmarks Post has posed against. I wonder, idly, if all the image making bores him. There's a round of thanks on both sides, and Andreas and his crew pile their equipment into their van and begin their drive home.

In the early evening, Post and I ride bicycles across the countryside to his home in a nearby village, passing farmland as we go. He points out badger holes and notes which of his neighbors raise pigs. I'm impressed with the stream of conversation he maintains as we park our bikes at the large house he shares with his wife, Liesbeth, and their two teenagers. Part of the house is a barn that Post restored himself. Carpentry is one of his hobbies, as I learned when one of his postdoctoral fellows mentioned he was thinking of buying a fixer-upper in the neighborhood and Post immediately offered the loan of his tools. Liesbeth has generously made dinner for the five of us, and we sit over bowls of pasta with pesto and bacon and reminisce about New England, where Liesbeth and Post lived for many years and where I was born and raised myself. Conversation quickly moves back to the lab, and Post reflects that the openness of his work on cultured meat may soon have to change. Venture capital funding will mean the need to protect intellectual property, which means inviting fewer visitors such as myself to the lab. We talk about ways of defining "food," and Post offers a definition appropriate for a man who's eaten the same sandwich each day for years: food is fuel. Later, as the evening wears on, I offer thanks for the fuel, and cycle back to Maastricht through the darkening countryside, grateful for rental bikes that come equipped with friction-powered lights.

The next morning is my last in the lab, and in the afternoon Post and I pile into his station wagon and drive toward Nijmegen. Our destination is a suburban conference venue built out of a medieval monastery, where Post will address a group of medical students. I ask him questions as we go. I've heard many observers opine that cultured meat is likely to progress more slowly than regenerative medicine, because medical research will attract more funding, and I ask him his opinion. He acknowledges that the technology for cultured meat begins in medical tissue engineering, but he suspects that regenerative medicine will progress behind a vigorous young cultured meat industry, absorbing techniques from the latter. We pass a pig farm and I ask Post why he chose to focus on cows. He answers that it reflects his prioritization of environmental and food security concerns, cows being the most environmentally damaging animals to raise, and the least efficient from the standpoint of feed conversion ratios. Chicken would be the most important target animal if animal welfare were his first priority. We raise far more chickens than cows, and arguably under far more inhumane conditions. When I press him on the issue, Post insists that he is not attached to the symbolism of beef in any way. His choice of beef has nothing to do with the hamburger's international prominence, nor with the prestige of steak. I ask him how he might respond to critics who don't believe cultured meat can work at scale, and he responds that the burden of proof is on the doubters. It's true that cultured meat represents an unprecedented effort to shift scales from the medical to the industrial, but Post sees no reason why a lack of precedent indicates impossibility.

Much later in the day, after he has given his talk and answered questions, when we are tired and on the road back to Maastricht, Post surprises me by mentioning Calvinism. Our conversation has drifted from Dutch incredulity at the privatized American health-care system, to different ways of drawing the distinction between science and engineering, to the current condition of the Dutch welfare state. Post brings up Calvinism in order to say something about his work ethic. Not the motivations for his work in cultured meat, per se, but why he, a scientist who has reached a secure place in his professional life, feels the need to resist complacency. In contemporary Dutch society it is easy, Post says, merely to sit with a glass of white wine and enjoy the day. He believes a climate of ease has settled over the Low Countries, and he fears the stagnation it brings. Science as a way of life has been Post's antidote to stagnation, and he has gotten so used to constant challenge that it is easier for him to charge ahead into the next project than to take pride in his accomplishments.

Post's invocation of Calvinism indexes a well-known history. Calvinism became an influential force in Dutch culture after the Protestant Reformation, combining a doctrine of predestined salvation with a doctrine of worldly activity. Material success, thought the Dutch Calvinists, supplied evidence of predestination for Heaven. Generations of this thinking built a culture of works and deeds, and according to one classical argument in the social sciences, this thinking contributed to the rise of modern capitalism itself. So argued the sociologist Max Weber in *The Protestant Ethic and the Spirit of Capitalism,* and as we roll back into Maastricht, I cannot shake the associations between Calvinist culture, modernization, and the perils of animal agriculture at massive scale. Cultured meat often seems to me like an effort to repair the damage done by modernization while working with the tools (and staying within the limits) of market capitalism itself. This would tie too neat a bow around the whole story, I know, but I can't help but think about the fact that many cultured meat entrepreneurs have described themselves as starting businesses in response to a moral imperative.

Months later, I am having dinner in a cave between Maastricht and the nearby city of Valkenburg, along with scores of scientists, journalists, and others. We are guests of Post on the occasion of the first annual Cultured Meat Symposium. While not the first international conference on the topic, it has a special importance as the first post-Post-burger conference, the first convened by Post, the first conference held at a moment when venture capitalists have become keenly interested in cultured meat. Post has assembled talks from specialists in tissue engineering, stem cell science, meat science, and food science more generally, and there is even a panel of social scientists, which I notice is marked in the conference program with a thumbs-up sign and the label "Acceptance."

The academics on the social science panel, incidentally, do not present data that fits neatly into a thumbs-up sign. Their work suggests a thumb in a wavering sideways position, perhaps indicating "Ambivalence." Survey research by Wim Verbeke examines the way respondents navigate the question of risk, not to mention feelings of disgust and concerns about the "unnatural" origins of cultured meat. The bioethicists Cor van der Weele and Clemens Driessen suggest that the ambivalence they have found marks a tension between the enjoyment of carnivory and concern for the welfare of animals.[10] Neil Stephens, the sociologist who coined the phrase "as yet undefined ontological object" to describe meat grown in vitro, states that said meat seems to have moved past its initial phase of ambiguity. That is, there is more and more consensus regarding in vitro meat's nature, a consensus captured by the term "cultured meat" itself. Although only some of the social science discussed has to do with early-stage surveys of potential cultured meat consumers, it is during these talks that the entrepreneurs in the room take out their smartphones to take pictures of the slides. "They'll take this back to their venture capitalists," whispers one science journalist with whom I've been chatting, "and fold it in to their sales pitches, as evidence of a market."

Such cynicism, I think, looking at the rock walls of the cave. They are made of "marl," a loose geological term for a combination of clay, silt, and lime or calcium carbonate. Within Sint Pietersberg, chalk and marl extraction has produced a series of galleries so extensive that some ten thousand inhabitants of Maastricht could hide in them while the Nazis, and the sympathetic Dutch who aided them, were in power.[11] Downed Allied pilots took refuge in the tunnels and were led underground, along what was called the Pilot's Line, to safety in Belgium. Rembrandt's painting *The Night Watch* masqueraded as a stalagmite, safely rolled up until Maastricht was liberated in September 1944. But the cave we are in now is not in Sint Pietersberg; we're a bus ride away from the university, and during that ride my science journalist friend told me that the basic structure of most articles in science journalism gestures toward the future: introductory paragraph, description of a recent breakthrough, then informed speculation about how it might change our lives. Perhaps, I think to myself, this affinity between science journalism and casual or "lay" futurism is where sloppy promises, or at least an ambient climate of informal promise, come from. In his opening remarks for the symposium, Post reminded us that our work gets media attention galore, and that we need critical debate within our community as well.

My belly is not full of cultured meat, of course; my five months' absence from Maastricht was technically enough time to make a burger, but we are not eating any cell-cultured Blanc-Bleu Belge tonight. The conference is packed with talks, many of them highlighting the connections between medical stem cell science and meat production. These, too, are largely speculative, given by muscle and stem cell researchers who believe that their work potentially has bearing on the production of cultured meat. The granularity of detail is impressive, from stem cell function to chemical engineer Marianne Ellis's observation, in the course of a talk on bioreactors, that there is a lot of wasted space in each and every T-flask used in cell culture experiments. Other talks examine the present state of industrial meat production and regulation in Europe, or the task of crafting life-cycle analyses for cultured meat. Isha Datar speaks on New Harvest's effort to promote cultured meat, among other products of cellular agriculture.

The first keynote speaker of the conference, Michael Rudnicki of the Ottawa Hospital Research Institute, is a stem cell specialist, and he speaks on the challenges of understanding the molecular mechanisms that control muscle stem cell function during the growth and regeneration of skeletal muscle. Normally quiescent muscle stem cells, Rudnicki explains, enter their cell cycle in a reaction to injury or the stress of bearing weight; this is what Anon triggered when he cut into the sample of meat for Deutsche Welle's camera. They then go through a process called asymmetric cell division to produce daughter cells, committed to becoming myogenic, or part of the muscle tissue, and these in turn produce more myogenic progenitor cells. At this point, other processes lead the initial stem cells back into quiescence.

Rudnicki's research has targeted muscle regeneration, which occurs in response to trauma or diseases such as muscular dystrophy, and seeks to address it at the level of molecular mechanisms. He has been investigating diminished stem cell function, and in particular the diminishment of the symmetric—in the sense of generating two identical stem cells—expansion of stem cells, which can replenish the initial quiescent stem cells on hand. Ordinarily, symmetric and asymmetric processes operate in a kind of balance, governed by feedback within the healing tissue, to ensure that muscle not only regenerates but also retains the capacity to regenerate after a subsequent round of damage or stress. Rudnicki has found that by adding a specific protein, he can increase the symmetric expansion of stem cells during the healing process. The medical implications concern not only muscle disorders and diseases, but also the diminishment of stem cell function due to aging.

Rudnicki's work interests this room of listeners because of the implications for meat production of coaxing muscle growth out of stem cells. When asked about a timeline for therapeutic or food applications of his research, which currently uses mouse rather than human models, Rudnicki declines to offer one. Always respecting this choice, I find myself musing about the meaning of repair giving rise to production, and about whether or not there might be limits to the ability of a healing process to "scale."

As we hear more papers on stem cell science, there are murmured objections and doubts. Someone does a little napkin math and tells me that if anyone tries to build a stainless steel bioreactor the size of an Olympic swimming pool, they will quickly find that there is not enough stainless steel in the world to accommodate their dreams. I have doubts about the accuracy of this complaint. The top of New York's Chrysler Building is clad in stainless, and so is St. Louis's Gateway Arch. Another critical observation cuts deeper. My science journalist friend points out that if there were bioreactor capacity to produce cultured meat, it would raise bioethical questions about the use to which those bioreactors should be put. Any bioreactor that could produce meat fit for human consumption could also produce human tissue fit for transplantation into human patients. Why produce a hamburger, which is a gastronomic mayfly creation that will not last a day, if you can produce cardiac tissue to save the life of a heart patient? From both moral and economic standpoints, there are simply no decisions to make here. If this is a purely hypothetical point, because it posits the existence of an almighty office of resource allocation free to choose between food and medicine, it nevertheless captures one dimension of the imaginative exercise of thinking about cultured meat. To think that cultured meat will be viable is to imagine that medical tissue-culture techniques can produce food at industrial scale without drawing resources away from medicine. For some, it is also to imagine that the moral and practical effects of a successful cultured meat industry might be, in environmental and moral terms, more valuable than medical advances. I have also met scientists and entrepreneurs who imagine that a rising tide of tissue-engineering science will lift all boats, ushering in a cornucopian era in which our capacity to grow tissue for medical and food applications effectively changes our relationship with living material, including our own bodies, altogether.

Post's express wish is that we found a scientific society to promote research into cultured meat, and he devotes a session to discussing what such an organization might be like. The result is a freewheeling argument between scientists and entrepreneurs, all of whom have their own ideas about strategy. We need to produce hamburgers, sausages, and other mimetic meat forms, one attendee says. No, says another, we need to create things consumers have never seen before. Should we explore existing regulatory pathways for meat in the European Union and the United States, or try to find a way around them? This, too, triggers a debate. Post reiterates that he was thinking of a scientific organization devoted to tissue-engineering techniques for the production of meat, and not a public-facing body whose role is to promote products and steer them through or around regulatory infrastructures and into customers' lives. However, even as he says this, the line between scientific practice and business practice seems to evaporate. How crucial is it, someone asks, to completely replace the meat industry as it stands? I am reminded that this room includes people whose interest in cultured meat is that of animal protection activists, as well as others for whom cultured meat represents a distinct market opportunity.

Someone else makes the excellent point that the division of labor between biologists and engineers shouldn't be, sequentially speaking, that the biologists work out a program and the engineers are left to implement it—in other words, biologists shouldn't leave the challenge of upscaling for other people to deal with. And very large bioreactors present engineering challenges that range from temperature control, to the circulation of growth medium, to the relatively even distribution of cells throughout the bioreactor chamber—a list that barely scratches the surface. More desirable than disconnected communication, according to the commenter, would be a running conversation between the two types of specialists throughout the development of a process for cultured meat production.

During a coffee break, one of the servers asks me if I've had any of the ice cream made with polar bear stem cells. Polar bear stem cells? I later find out it is an elaborate joke; she's referencing an exhibit put on by the Next Nature Network, the same organization that made the *In Vitro Meat Cookbook,* in which they present speculative foods of the future as if they were already here. Polar bears and ice and water (not to mention vanishing glaciers) are on my mind as I plunge back into the conference. Meat, I hear a speaker say, is mostly structured water, which is true enough: 75 percent water to 20 percent protein to 5 percent fat is one ratio I often see used to describe meat's constitutive elements. Datar says that we can't yet imagine how meat will eventually be defined, when we have started to grow it in labs, and I find myself nodding vigorously. What, after all, is the shape of water? At the end of the session, Post rises and puts his hands together to applaud, and so do we all.

FOURTEEN

Kosher

Jews have been asking if artificial flesh might be kosher for a very long time. The question even appears in the Talmud (the collection of rabbinic commentary on the Hebrew Bible). In the Talmudic Tractate Sanhedrin (65b), two rabbis, Chanina and Oshaia, spend every Sabbath evening studying the Kabbalistic Book of Creation, the *Sefer Yetzirah*. They use its teachings to create a calf, which they then eat without slaughtering it in accordance with *kashrut* (kosher law). Despite this transgression, interpreters of the tractate have had reason to wonder whether Chanina and Oshaia actually violated *kashrut*. The manmade calf may not have been a "real animal," at least according to Rabbi Yeshaya Halevi Horowitz (active during the late sixteenth through the early seventeenth century). Its unnatural status meant that *shechitah* (the practice of kosher slaughter) was unnecessary. But other authorities, though they agreed that the calf was unnatural, opined that the failure to perform *shechitah* violated a different principle, that of *marit ayin:* the avoidance of actions that might appear improper even though they are not actually improper. Thus, even an artificial calf would have to be slaughtered appropriately, lest Chanina and Oshaia appear to hold themselves above the law, consuming a seemingly natural but unkosher, because not properly slaughtered, calf. This is a story about keeping up appearances, but also about the treatment of animals, and the relationship between meat and the cycle of animal life.

In another Talmudic narrative (Tractate Sanhedrin 59b), a different kind of artificial flesh appears. A traveling rabbi, threatened by lions on the road, prays to the heavens and receives an immediate answer: two hunks of flesh fall from the sky to distract his assailants. Obligingly, the lions seize one hunk, leaving the rabbi free to carry the other to a study hall. There the piece of meat feeds a different kind of hunger: a hunger for debate. The final judgment on the meat is that "no unfit thing descends from heaven." The meat is not only "good for the Jews" but fuels both the body and the mind. This story, like that of Chanina and Oshaia, invites us to wonder where miraculous or magical meat might fit in the order of nature. What mental operations are needed to place a new form of meat in our picture of the world, and within the regulatory system that governs our eating?

The laws of *kashrut* are ancient and modern. The original meaning of the term *kasher* (kosher) is "fit," in the sense of "suitable." The Talmud, which was fully compiled in its Babylonian form by about the fifth century C.E., nevertheless considerably postdates the Biblical origins of *kashrut* in the books of Leviticus and Deuteronomy. Meanwhile, the modernity of the laws of *kashrut* comes from the way they change in response to shifts in how Jews produce and consume their food. Over the past hundred years, and like all other eaters in the developed world, Jews' dietary patterns have been transformed by food's industrialization. Everything from gelatin to white bread to meat, not to mention nonfood household products such as hand lotion and shampoo, can be bought with a kosher label on it, all produced at factory scale. Many of those labels, placed by a specific body that oversees kosher certification, reflect debate and argumentation over the kosher status of specific ingredients, debate that has itself been an important part of the process of Jewish modernization. Now the rolling debates over *kashrut* have found laboratory-grown meat.

In the late spring of 2016, I was in the audience when a representative from New Harvest spoke as part of a panel discussion on the future of food, in Oakland, California. When I asked if cultured meat would be kosher, she happily reported that the issue of *kashrut* had recently been settled during an Internet discussion in which one contributor cited a rabbi's suggestion that laboratory-grown meat would pass muster.[1] Thus, my interlocutor continued, there would be no reason for Jews not to participate in the future of protein. Despite this confident answer, the potential kosher status of cultured meat products remains ambiguous, and the rabbinic community has been polyvocal rather than unanimous.

Another organization that promotes meat produced via tissue culture, the Good Food Institute, has (as of 2016) advocated that we replace "cultured meat" with another term: "clean meat," which is remarkable for its ability to make the Book of Leviticus echo strongly into the twenty-first century. As is so often the case, the term "clean" here summons its opposite, and the implication is that conventional meat is "dirty." Thus, the term "clean meat" has angered many farmers and meat industry lobbyists. The anthropologist Mary Douglas writes, "Defilement is never an isolated event. It cannot occur except in view of a systematic ordering of ideas," and this is true of cleanliness as well.[2] Viewed through Douglas's lens, *kashrut* is just such a "systematic ordering of ideas," a way of structuring human experience and behavior. Douglas's *Purity and Danger* was in my backpack when, many years ago, I found myself at a "pig-picking" in rural Kentucky, pulling meltingly soft meat off a hog's flank with my hands as steam rose from the apples that had been roasting in its gut. The animal had spent hours in a pit in the ground, surrounded by hot rocks. Its meat seemed clean to me, and, if memory serves, it was delicious.

The rabbi cited in New Harvest's Internet discussion opined that, since no animal need be killed to produce cultured meat, the meat would not count as meat at all from the perspective of *kashrut*. Cultured meat might be considered *pareve*, the category for foods that are neither *fleischig* (meat) nor *milchig* (dairy), the two categories that are never to be mixed, according to kosher law. The first response within the same Internet discussion thread—notably offered without a trace of anti-Semitism—was a heretical but high-spirited form of "bacon glee," a series of jocular comments about bacon and pork, beginning with the suggestion that, if meat grown from cow cells need not count as meat at all, why not grow "kosher" pig cells? The transgression continued, spinning up to full-on Jewish Humor mode: "Kosher bacon, the dream is real." "We might make that our next fundraising campaign." "If you make kosher bacon you need to film a Rabbi eating it." A few years earlier, someone had created a website in order to tell a different kind of transgressive joke. They claimed to have a company that could create cultured meat out of the muscle cells of celebrities. This was less a suggestion that we should practice cannibalism than a second-order joke about getting to fulfill a hidden wish: to consume the stars whom the tabloids insist we follow, carrying both our desire to possess their lives and our resentment of their prominence to their natural, if socially taboo, conclusion.

New Harvest was not the first to raise the question of the kosher status of meat produced via tissue culture techniques. Curious inquiries began to pile up on the Internet around August 2013, when Mark Post unveiled his famous hamburger made from cultured cells. Early conversations about cultured meat and *kashrut* have often focused on the issue of which animals are permitted or forbidden to Jews, as described in Leviticus 11. In other words, the first questions assumed that the source animal for a cell culture would be one of the major determinants of *kashrut*. On this reasoning a biopsy from a cow might lead to a kosher cell line, but a biopsy from a pig or a camel would lead to a forbidden cell line.

However, and as the Talmudic story of the rabbis Chanina and Oshaia and their artificial calf shows, kosher status depends on more factors than just animal type. First, sick or injured animals of any species are unkosher. Second, even fit animals of the proper species are not considered kosher unless slaughtered correctly according to the rules of *shechitah*. Animals killed this way are not stunned before slaughter, as is the nonkosher convention. They must be awake and alert as a long ritual knife, called the *chalif,* is pulled across their throat, instantly killing the animal while beginning the important process of draining it of blood; some even argue that this method of slaughter swiftly deprives the brain of oxygen, minimizing suffering.[3] After the body has been drained, the carcass is carefully inspected for deformities, growths, ruptured veins and blood vessels, stagnant blood, and so forth—all in conformity with the Biblical injunction against eating blood or a diseased animal. The blood ban has sometimes meant that many kosher-keeping Jews eat meat only from the forequarters of an animal, rather than from the hindquarters, the veins and arteries of which are very tough to remove. The sciatic nerve is also considered unkosher. The process is, of course, labor intensive, and as a result kosher meat usually costs significantly more than its nonkosher equivalent, another reason why it is, as the Yiddish saying goes, "shver tzu zein a Yid"—hard to be a Jew. Nor is the rabbinic community unified on the question of whether or not kosher meat is, itself, a more ethical option than nonkosher meat. One rabbi wrote, in a 2014 newspaper editorial, that because the vast majority of kosher meat production took place under conventional industrial conditions, it was in fact no better than its nonkosher counterpart.[4]

Notably, and very significantly for the production of cultured meat, according to the principle of *aver min hachai* (a limb from a living animal), one cannot sever and eat part of an animal while the animal still lives. This largely depends on whether a rabbi judges that a biopsy of tissue, taken from a donor animal, does or does not count as a "part" of that animal—and would a hamburger grown from that sample also count? To kill the animal in order to eat its cultured cells would displease many celebrants of cultured meat, even if (theoretically) a single biopsy might yield tons of meat via cell culture; a sacrifice of a specific animal could, given an affirmative rabbinic ruling, yield many animals' worth of kosher flesh.

Aver min hachai leads to a Jewish version of a more general, and philosophically enticing, question about cultured meat. What is the relationship between the tissue culture made from an animal and that animal itself? Does a muscle fiber grown from a cow's cells share an essence with that cow? If so, is that essence somehow borne by the DNA in the cells in a biopsy? Should we understand the relationship between donor animals and their cultured meat on the model of parenthood, or as an extension of an original adult body that may or may not still be alive? Where, for that matter, are the boundaries of the animal body? Perhaps more practically, if cultured meat were judged to be *pareve* rather than *fleischig,* what would that judgment mean for the broader question of the "meatiness" of cultured meat? And how much would such a Jewish opinion weigh, in a broader (that is, including both Jews and non-Jews) conversation about whether tissue-cultured cells could count as meat, in all the conventional senses? It is easy to imagine that if a religious authority judged cultured meat to be kosher, but only on the grounds that it was not actually meat, cultured meat's advocates would have to think twice before agreeing. For them, this might be the wrong "systematic ordering of ideas."

In 2016, a new entrant into the race to create cultured meat, the Israeli start-up SuperMeat, made much of the issue of *kashrut.* As journalist Sarah Zhang pointed out in the *Atlantic,* company cofounder Koby Barak was transparent about the controversial status of kosher claims, particularly ones based on still nascent technologies.[5] Kosher claims on behalf of industrial products must be based on determinations about individual ingredients.[6] In the case of cultured meat, the growth medium used to feed the cells, the scaffolding on which they grew, and the original cells themselves would all require inspection. Conscious of the problems that may arise from a nonkosher source ingredient, the rabbis consulted by SuperMeat seemed ready to turn to an interpretation of one particular kosher principle, known as *panim chadashot* (a new face). *Panim chadashot* means that if a substance's physical form is radically transformed, then an originally unkosher ingredient may lose that status and be considered kosher in its new form. In one case, kosher status was granted to gelatin derived from the collagen in pig skin on the grounds that the ultimate gelatin seemed to consist of a completely different substance than the animal product at the beginning of the process. But the application of *panim chadashot* did not go undisputed. The pig gelatin case led to scandal despite the affirmative rabbinic judgment, and producers of kosher gelatin ultimately had to switch to other sources of collagen. This specific case may hold implications for cultured meat, since collagen is a useful ingredient for making the organic scaffolding on which tissue cultures are grown.

The story of *kashrut's* encounter with industrial food production is a complex one. At its center is a question central to the modern Jewish experience, namely, how can observant Jews participate in the non-Jewish world without transgressing the limits of Jewish law? That question would take on a distinctive form in American life: could assimilation, in the sense of life within the mainstream of American society and culture, take place on Judaism's own terms?[7] More prosaically, could one drink Coca-Cola made according to the original recipe, which included a small amount of glycerin and thus was not necessarily kosher? Could an observant Jew only drink Coca-Cola made with a kosher form of glycerin? Of course, Jews have many centuries of experience adapting *halacha,* the legal system of rabbinic Judaism, to the customs and resources of the nations in which they live. The issue of *davar hamamid* (that which makes stand) is instructive: ordinarily, according to the principle of *bitul* (nullification), the presence of a nonkosher element in a kosher food or liquid can be negated if the ratio between the two is 1:60 or greater. However, in certain cases the small nonkosher element may play a catalytic or determining role for the overall structure of a thing: "making it stand," as *davar hamamid* implies. A classic premodern example is the use of animal rennet (typically taken from the lining of a calf's stomach) to turn milk into cheese. In a modern context, a great many industrial foods employ such catalysts, whose kosher or nonkosher nature needs to be ascertained.

The history of *kashrut* in the twentieth century is also one of shifting regulatory regimes and shifting markets, and of a certain "interpretive pluralism," the presence of multiple bodies making certification decisions that will be received by specific communities within the Jewish world. Orthodox, Conservative, and Reform Jews all tend to interpret *kashrut* differently, and different kosher authorities produce different brands of kosher certification. Consumers make their determinations about which kosher labels, or *hechshers,* they will trust, which often but not always is determined by their denomination within Judaism. Beyond this, there are federal laws that regulate the kinds of representations koshering agencies and product manufacturers may make, in addition to governing the private litigation to which businesses may have to resort. The result is a hybrid public-private regulatory environment for kosher products, and one in which business interests play an important role. The organizations providing kosher certification get paid for this service, while the food producers whose products they certify receive access to specific kosher markets thereby.

Perhaps it is not so strange that the question of whether cultured meat might be "fit," or *kasher,* has appeared preemptively. It reflects the early stirrings of a regulatory imagination, because we onlookers (Jewish and non-Jewish alike) know that new food products will be tested and labeled by government agencies before they can be sold. Early debates over cultured meat's kosher status begin to seem like a rehearsal for the later debates that will affect whether or not, or under what conditions, cultured meat can be sold. The Talmudic stories put this matter plainly when they ask under what category we may place meat that has no clear or natural animal origin. Cultured meat may not slide easily into our "systematic ordering of ideas" about the relation between meat and animal bodies. It could force us to rethink that ordering or abandon it entirely. We may find that the parts of the world no longer fit together for us, when meat does not point back toward the larger shape of the animal body.

FIFTEEN

———

Whale

How did this conversation get to whales? I'm in an auditorium at the Business School at Stanford University. Paul Shapiro, vice president of policy for the Humane Society of the United States, has the last word at a panel discussion about cellular agriculture and cultured meat. Over the past hour and a half, the large audience gathered here has heard from Shapiro, as well as from Bruce Friedrich, founder and director of the Good Food Institute, which, like New Harvest, promotes cellular agriculture as an alternative to animal agriculture. The Good Food Institute has sprung up almost out of nowhere over the past year, funded by deep pockets and strongly allied with animal protection organizations. Uma Valeti, founder and CEO of the cultured meat company Memphis Meats, has also spoken, as has Cor van der Weele, professor of philosophy at Wageningen University and one of the earliest bioethical commentators on cultured meat. It is evening, and portions of the crowd seem tired after a full day on a busy university campus, but others seem eager for the networking session that breaks out after the closing applause, taking their photographs together and exchanging business cards. This has not been a divisive meeting. There has been no debate about whether cultured meat, which Friedrich and Valeti both prefer to call "clean meat," is an appropriate strategy for securing protein sustainability, climate change mitigation, or animal protection. There has been no conversation about whether resources might be better used some other way. Any minor disagreements have been swept aside by a shared enthusiasm for a new technology and its capacity to create positive change.

Shapiro is the one with cetaceans on his mind. Good at getting laughter and applause from the crowd, he ends the discussion with a tale from the late days of American whaling. Just before the Civil War, he tells us, the whaling industry was America's fifth largest. Every home in the country had its lamps, and those lamps were fueled by whale oil. But between 1853 and 1873, the fleet of ships that hunted and killed those whales shrank by 80 percent, largely because of a new product, kerosene. The Canadian geologist Abraham Gesner discovered a process for deriving kerosene from petroleum, and this new product swept through the fuel market, displacing whale oil. A single technological development, in other words, led to an alternative to dependency on an animal product. It is a good story. Shapiro tells it to provide a precedent for laboratory-grown meat. I listen closely, because this way of using a specific historical case as a precedent for the future of technology is strikingly common in cultured meat circles. Moreover, this style of mining historical accounts is a feature of many public discussions of emerging technologies.[1] But a new technology is not the only "actor" in the story. The market acts too, and the better alternative wins as if through natural selection. It is an intriguing story for an animal advocate, as opposed to a technology merchant, to tell.[2] It suggests that he has adopted the methods of such merchants.

Shapiro's version of the story is very widespread.[3] You can even find it in natural history museums and museums devoted to whaling. The story of whale oil contains ambiguities, however, and it has been retold for other, highly politicized, reasons in other contexts. According to the environmental historian Bill Kovarik, a "whale oil myth" has often served as a centerpiece in arguments that technological innovation, combined with the free market, can save elements of the natural environment from the worst predations of industry and development.[4] It is true that technological developments introduced fuels that shifted Americans away from whale oil in the mid-nineteenth century. Yet the most widespread alternative fuels available when whale oil began its decline were not kerosene, but rather alcohol-based blends of various kinds, including a particularly popular one called camphene, made of alcohol and turpentine, which was substantially cheaper than whale oil; camphene's chief defect was its volatility. When kerosene appeared on the market as a lamp oil, whale oil (and thus the whaling industry) was already years past its peak. Kerosene overtook its alcoholic competitors not because of superior performance, nor as a result of its merchants' business acumen, but because of taxation. Kerosene's rise, and the rise of the petroleum industry at large, was aided and abetted by government taxes on alcohol during the Civil War; alcohol, whether for drinking or for burning in a lamp, was heavily taxed compared to the lighter levies placed on kerosene.

The happy ease of Shapiro's story belies the complexities of the case. A technological breakthrough, propelled by the market's natural preference for a better, cheaper product, did not intervene to save the whales. The story of whale oil's decline seems to point to neither a hagiography of invention nor a celebration of the market. But it does point directly to the importance of government activity in mediating the emergence of new technologies and the decline of established ones. No market, and no inventor, operates in a political vacuum. In 1830, some fifteen years before the whale oil market peaked, the whaling agent Charles W. Morgan delivered a lecture entitled "The Natural History of the Whale" in that great center of whaling, New Bedford, Massachusetts.[5] Morgan, who had a financial interest in a New Bedford workshop that made spermaceti candles, worried about a reduction of the import duties on olive oil, a competing product. He was keenly aware that whaling took place amid regulation and taxation, and alongside competing fuels that could cut into his profits. Morgan also spoke of "carbureted hydrogen gas," which he acknowledged had certain advantages over whale oil, for static lighting and for brightness of illumination, but which he thought was too "subtle" (i.e., volatile) a substance to travel well.

Is the case of whale oil applicable to cultured meat at all? There are reasons to doubt it. Whale oil became scarce because it was taken from slain whales faster than whales could breed and grow and breed again; industrial meat production may lag behind increasing demand in the developing world, but not in quite the same way.[6] Shapiro smiles as he describes an 1861 cartoon of cetaceans, which ran in *Vanity Fair:* whales cavort at a party, like human socialites, delighted by the oil now flowing from Pennsylvania soil, where a man named Edwin Drake had sunk his well. Drake's oil fields, dubbed "Petrolia," would soon supply most of the oil in the United States and, in time, most of the petroleum in the world. It is little wonder that kerosene lamps were taken to be the salvation of the seas. It is very easy to imagine a similar cartoon in which cows and chickens and pigs cavort, delighted that an artificial surrogate for their meat has been invented. This is the emotional heart of the story, the thing that motivates Shapiro's telling of it, entirely orthogonal to the intricacies of a regulated, and thus inevitably political, market into which a new fuel source splashes. All the rough spots of history get flattened out, and the story of the whale becomes a natural resource of sorts, a precedent for innovations to come.

SIXTEEN

———————

Cannibals

"Why don't we culture and eat our own cells?" The joker, a medical student, sits in the back row of a large auditorium where all eyes are on Mark Post, who is conducting a question-and-answer session following a lecture on cultured meat. Less a wave than a shudder of nervous laughter ripples the room. Some of us have heard this one before. During the storm of hype surrounding Post's hamburger demonstration, a website went up advertising a fictitious company that offered meat grown from the flesh of celebrities, and the idea of culturing and eating our own cells has circulated in the "comments" section of Internet news stories about cultured meat. The provocation of the joke seems shallow, but it isn't. After all, if cultured meat is (among other things) a romantic garden fantasy in which our appetite for meat is pacified, in which we are no longer killers, it makes sense to worry about the garden's nighttime side, when our monstrosity, submerged all day, rushes from below.

Post laughs quickly. He too has heard this one before. Post stops laughing and takes things further. "I get this question from kids aged eight to twelve pretty often," he says. "It makes sense." Post then takes a breath. I expect him to say something about how culturing and eating our own cells would be a weird translation of cultured meat's origins in medical tissue culture. Instead, Post gets psychoanalytic. Sigmund Freud tells us, he says, that the desire to taste ourselves is part of normal sexual development, a form of erotic fantasy we learn to suppress as we grow up. Children lack such adult limitations. More laughter. The medical student has become the butt of his own joke via Post's psychosexual interpretation, and he has the good grace to laugh along at his own expense. You might say his joke is "dirty," in the sense the anthropologist Mary Douglas once intended when she called dirt "matter out of place": our tissue could become matter out of place in the food chain, occupying a position previously filled by pigs or goats.[1]

In his essay "From the History of an Infantile Neurosis," Freud writes that the oral phase of our development has left "permanent marks" on our uses of language. Thus, we speak of objects of lust as "appetizing" or call our lovers "sweet."[2] Elsewhere Freud argues that in very young children the desire for food and the sexual drives (which do not suddenly emerge in early adolescence but exist throughout childhood) remain unified, not yet separating as they will when the child emerges from sexual latency.[3] Thus, sexual desire is comprehensible to the child only as the wish to consume desire's object, incorporating it into one's own body. Cannibalism is pregenital sexual organization. In a few scattered places in his writings, Freud refers interchangeably to the "oral" or "cannibal" phase, the latter being the term for anthropophagy that likely originated with Columbus's voyage to the West Indies. One of the peoples Columbus encountered called themselves the "Canibales," and he suspected them of eating human flesh. The word spread from Columbus through the languages of Europe, as Europeans looked to terrae incognitae and imagined that the humans in those lands engaged in the transgression of eating one another. For some Enlightenment thinkers, cannibalism even featured in thought experiments concerning the population capacity of far-off and isolated places, islands in particular. After how much reproduction, the hypothetical and fantastic inquiry ran, would anthropophagy have to begin in order to balance population and food supply?

Cannibalism was a common feature of the geographies early modern Europeans imagined. It was, you might say, a moral mark on those maps. And cannibalism had an ambivalent character in the eyes of many European writers who speculated upon it. Was cannibalism a violation of the natural order of human social life and subsistence strategies or was it, conversely and disturbingly, part of life in a state of nature?[4] Freud's late modern solution was to argue that it was both, and that (contra his early modern predecessors) the mistake was to imagine that cannibalism was something that happens somewhere else, among persons radically unlike ourselves, living outside our moral community. The impulse toward cannibalism, Freud wrote in his late work on religion, *The Future of an Illusion,* is "born anew with every child" along with the wish for incest and a lust for killing.[5] Of these instincts, Freud said, only cannibalism "seems to be universally proscribed and—to the non-psychoanalytic view—to have been completely surmounted."[6] Which, of course, is to say that according to a psychoanalytic view cannibalism has not so much been surmounted as contained through the repressions of civilization.[7]

Civilization, Freud argued, arises to protect us against the powers of nature both within us and without. Yet in a state of civilization the power of culture that defends us against nature also divides us from ourselves. In Freud's *Moses and Monotheism,* a band of brothers achieves a proto-civilized life under a rough social contract only after first killing and eating their tyrannical father.[8] In a world of cultured meat and sophisticated tissue engineering, the option to eat our own flesh or that of our species' kin is difficult to ignore entirely, which explains the persistence of nervous jokes about it. Freud's story about cannibalism may disturb those who wish to believe either that children are moral innocents or that cannibalistic impulses arise only as a disorder in the human soul.

The idea of growing human meat in a lab may seem first and foremost like a psychological provocation, but it is an anthropological one too, because it forces us to ask what we are, as humans, if we can also be a type of meat. This would be flesh that has never been part of a complete human body. A new form of human life reduced to sheer cellular metabolism. Perhaps what is truly unsettling about the concept of cannibal tissue culture experiments is not that we might eat one another, or ourselves, but that technology might introduce a new plasticity into our concept of what it is to be human. The flesh in the bioreactor is not sleeping; we are not waiting for it to wake up and be freshly animated by human will. Our cells, grown in tissue culture, implicate us in the order of livestock, and to eat them would mean embracing this reordering of the human condition.

SEVENTEEN

Gathering/Parting

"That makes me think of a diagram. Have you read Lévi-Strauss's *Elementary Structures of Kinship?*" Jordan seems excited by an association he's having. I've been telling him about the first day of New Harvest 2017, the cellular agriculture conference that has brought me back to New York City. My friend and host opens his laptop and shows me a line drawing of a *mithun* or *gayal,* a domesticated bovine common to northern Burma, called a "buffalo" in this translation of Claude Lévi-Strauss's book.[1] The diagram illustrates how a *gayal* would be carved up and its parts divided among family members during certain ceremonies of the Zahau Chins. Lévi-Strauss writes that "the methods for distributing meat in this part of the world are no less ingenious than for the distribution of women." A stunning phrase when taken out of context, it suggests a wedding ceremony and a bride-price counted out in bovines. Lévi-Strauss is drawing from a 1937 article by H. N. C. Stevenson, "Feasting and Meat Division among the Zahau Chins of Burma," which describes the way ceremonial meat division provides a map of kin relations under the local climatic conditions of patriarchy.

Best not to read too literally. Lévi-Strauss's interest here is neither meat nor women but the character of the rule-governed society, which uses rules to respond to the survival problems of scarce food and scarce reproductive opportunities.[2] He describes an early twentieth-century society that has not yet been threatened by the modern structural problems of cheap meat and overpopulation. Social rules began, Lévi-Strauss says, with the ur-rule of the prohibition against incest, which one might call the original form of social control over reproduction, one of the ways we "domesticated" ourselves as a species. But once they started to proliferate, rules became the system of symbols through which a society muddles along. Culture's vernacular begins with a ruled grammar. Over time this grammar proves its worth through intergenerational continuity. In the shorter timeframe of individual lives, ceremonial acts give shape and form and moral weight to social relations. In the case of the Zahau Chins, meat distribution binds those who give and receive meat in a network of reciprocal obligation. The *gayal* diagram is less for carving than for relating. Social bonds are "carved at the joints," as the expression goes, just like the recently sacrificed *gayal*. The diagram implies a wordless isonomy between community, marriage, and animal: whole beast, whole band—have joints, be joined. Jordan's association to Lévi-Strauss tilts my immediate experience of cultured meat on its axis. What on earth would it be like to share meat this way? To imagine animal bodies as the medium through which I feel the bonds of reciprocal aid? And could in vitro meat transmit the same feelings as in vivo meat?

I spent much of yesterday at the conference, looking at very different images and diagrams from the world of cultured meat and cellular agriculture research, many of them gesturing toward a world in which no innocent *gayal* need die in order for reciprocal kin relations to be ceremonially secured. I also spent my day thinking about different kinds of reciprocal obligations, including my own debts to the people who have shared their work with me over the past four years. I feel keenly my lack of a synoptic summation of all that I have seen and heard, and there are still no certain predictions or forecasts to be made on behalf of cultured meat. No intellectually responsible futurist would offer one. Nevertheless, a confident and easily consumed prediction is all that people seem to want from me. Bullet points. Narrative wrap-up. Here the scholar's target, which is to reach a new understanding, is nowhere near the goal of the public conversation. I want to leave more questions than answers on the page rather than impose a narrative line upon disparate images.[3] But if I am honest, I have also grown weary of other people imposing their own narrative lines through press releases and breathless journalistic celebrations. They wish away the obscuring clouds that surround any emerging technology.

During the first session of the conference, New Harvest's research fellows reported on experimental research toward cultured meat. All the speakers in this session were women, and the audience response was both surprised and positive; we've all grown used to the gender imbalance of the tech world, not to mention tales of widespread sexual harassment in tech and in academia, and the women-led New Harvest offered us an image of a different future. On the projection screen behind the stage the researchers displayed pictures of cells grown under culture, using laser pointers to indicate the long structures of muscle fibers. Or they wowed the crowd with images of a spoonful of turkey muscle cells, or showed images of sponges and other biocompatible materials to tell a story of how muscle cells can be anchored in tissue flasks to ensure proper growth. Much of the research presented was related to the two primary roadblocks in cultured meat research, which are the same ones that loomed in 2013: the need for a serum-free and sufficiently inexpensive growth medium, and the challenge of producing three-dimensional or "thick" tissue, which usually demands a sophisticated bioreactor featuring vasculature of some kind. We badly need to clear these roadblocks if cultured meat is to become a marketplace reality.

Around three hundred of us gathered at Pioneer Works, a very large brick event space on the outskirts of the Brooklyn neighborhood called Red Hook, and we sat on folding chairs in deep rows surrounding the stage. My mind wandering as it does, I imagined the building—once the home of Pioneer Iron Works, which made boilers, crushers, and engines for the Cuban sugar industry in the late nineteenth century—filled floor-to-ceiling with the kinds of tanks you see in beer breweries. "Pioneer Meat Works" would be a "carnery" producing pork and beef and chicken for a hungry New York.[4] Thinking of the building's historical use, I recalled that sugar plantations were once part of a frontier of resource extraction for Europeans in the Caribbean, and that one of the odd things about cultured meat is that it proposes we develop a new interior "frontier" in spaces we already occupy. In this frontier, cells themselves become a new agricultural substrate. Marianne Ellis, a professor of chemical engineering at Bath University, followed the research fellows onto the stage. In her talk she displayed a more industrial alternative to the carnery: a very complex diagram for a cultured meat production process linking raw materials to bioreactors to tissue engineering stations. This too was a kinship diagram of an indirect kind, illustrating the relationship between parts produced by diverse members of a very large team, which might include some of Ellis's own students.

Ellis had cofounded a small cultured meat company with a genial and well-spoken Welsh pig farmer named Illtud Dunsford, whom I first met at Mark Post's Maastricht conference, a rare visitor from the world of actually existing, *terroir*-dependent small-farm agriculture. Ellis and Dunsford planned to produce cultured pork based on cells from a heritage breed of pig that Dunsford raised on his farm. This was one of the few extant cultured meat projects that would copy not a popular form of commodity meat, but rather a kind of specialty meat. I made a mental note of the tension between the desire to replace that behemoth, commodity meat, and the small-scale, nimble profiles of start-ups like Ellis and Dunsford's, which seemed to reach for a niche market. To Ellis's great credit, when asked "How close are we?" during the question-and-answer session, she answered that she simply doesn't know. The same question was asked of Post in 2013 and repeatedly thereafter. It was asked by PETA in their 2008 chicken nugget contest. Winston Churchill's "Fifty Years Hence"—an article published in 1932, in which he speculated on the social and technological changes, including in food production, that science might achieve by the 1980s—merely gestured at it. This is the question we have been trying to answer for years, and though it has become boring, it is still on our lips. In the language of *Star Trek,* a familiar science fiction touchstone for many of the people around me at New Harvest 2017, "How close are we?" is the Kobayashi Maru, an unwinnable training scenario that teaches future starship captains that some losses are inevitable. Ellis's economical "I don't know" dismisses the question as an unfair trick.

Many of the research fellows' slides showed data from collaborative research funded by New Harvest, which in turn gets most of its money in small contributions from individual donors. The fellows make good on the investment made in them by participating in New Harvest programs, by sharing their research, and by publishing their papers on an open-access basis—but most of all by making progress toward a world of cultured meat and other products of cellular agriculture, thereby reducing or removing the need for animal agriculture. "Animal products are delicious," Isha Datar said in her opening remarks, but she also described their deadly effects on the environment when produced at industrial scale: global animal agriculture releases some 7.1 gigatons of carbon into the atmosphere per annum, by some estimates a whopping 14–18 percent of our total global carbon emissions. Bovine enteric methane alone releases 2.7 gigatons. The differing research backgrounds and goals of the New Harvest fellows are a source of strength, as I learned by occasionally attending their meetings. They share immortalized cell lines, which are able to divide and proliferate indefinitely, either through mutation or by dint of artificial intervention, such as the application of viral DNA. New Harvest has expressed interest in establishing cell lines that could be used by any researchers engaged in open-access cultured meat research. More immediately, cell lineages represent lines of genetic continuity linking cultured meat laboratories, shadowing the informal lineage relations that bind the community of cultured meat workers.

Lineage lines may be visible only to a few intimate observers, but one social distinction is very clear to all. The New Harvest fellows work in academic laboratories, advance through doctoral programs, and open their research findings to all eyes. They are part of a movement in scientific research toward open-access publication and information sharing, prominent examples of which include PLoS (Public Library of Science) and affiliations of serious hobbyists working in "hackerspaces." At this conference, "biohacking" was represented by Yuki Hanyu of Shojin Meats, which connects serious hobbyists across Japan. Yuki has used a sports drink as a growth medium with which to feed cells. He has grown muscle cells using materials that are easy to find at Tokyo *konbini* (convenience stores), and he proved it here with pictures. Shojin Meats takes its name from *shojin ryori,* the vegetarian meals traditionally eaten by Japanese Buddhist monks. Turning his laptop around on his display table, Yuki eagerly showed me a cultured meat–centric manga created by a Shojin Meats artist still in high school, in which carneries make meat in tanks on other worlds. He tells me that all this is inspired by a problem endemic in Japan, namely the fact that the island nation cannot grow enough food to support itself independently of imports. In contrast to this evidence of open work and play, the employees of cultured meat start-ups work in closed laboratories and their labor generates intellectual property for their company and its investors. Information about their progress stays in its black box until a product appears. For some, this is a very worthwhile price to pay for large amounts of venture funding.

The divide between academic researchers and their counterparts in enterprise is one that New Harvest did not address at the conference, which seems both deliberate and wise. There is a strong consensus that collaboration between academic and industry researchers serves the best interests of cellular agriculture, and New Harvest's leadership is eager to avoid any appearance that they side with academic researchers over start-ups. As they indeed do not. Datar has helped to found at least two companies, Perfect Day (formerly Muufri), which makes milk, and Clara Foods, which makes egg proteins. But because New Harvest devotes its resources to supporting open-access academic research and the training of new scientists, they are invested in painfully slow growth, from the perspective of the start-up economy. Datar did give the start-ups one important shout-out in her introductory remarks: Memphis Meats had brought the lab costs of producing chicken down to $9,000 a pound, which may seem astronomical but is vastly less than the 2013 cost of producing Post's first burgers. This rang in our ears as progress. The same company reported a 2016 cost of $2,400 to produce a meatball.

Journalists at the conference observed the divide between industry and academic research and asked me which group I believed. This casting of everything in terms of trustworthiness disoriented me. But their question was sensible. We were watching presentations about early-stage cultured meat research in the morning, chatting over a catered vegan lunch at midday, hearing the promises of cultured meat start-ups in the afternoon, and gathering for critical debate and beer in the evening. As of late 2017, the start-ups' promises were starting to merge into what the futurist Peter Schwartz calls an "official future." This is a rallying point for a diverse crew of actors who share assumptions about future events, which helps orient their actions in the present.[5] An informal affinity group has formed between several of the cultured meat start-ups and some of the animal protection and advocacy organizations that have taken an interest in cultured meat, especially the Good Food Institute (GFI), which exists to promote alternatives to meat. This affinity group takes the position that cultured or "clean" meat (to use the GFI-approved term) is feasible and imminent, and that imminent products may mean a faster demise for animal agriculture. As I interpret the reports of the research fellows, such a triumphal blowing of horns on behalf of start-up meat is premature, but I also know that opinion among the fellows is divided. Some take the start-ups' official future more seriously than others. Still, the shortfall between the public celebration of imminent "clean" meat and the apparent data (at least the openly available data) was obvious to everyone with whom I spoke on day one of the conference. Seemingly out of general politeness and a sense that everyone there was ultimately on the same side, no one made any public statement to draw attention to the issue.

New Harvest insists on transparency, and their response to the problem of hype is to move quickly toward higher ground. Datar remarked that she'd like to see a formula for animal-free growth medium circulate as freely as the cookie recipe on a package of chocolate chips. At another conference, New Harvest made transparency literal by setting up a display of empty plastic cell-culture flasks equal in number to the ones Post's team had used to make their burgers. The result was so much translucent plastic that I thought of modern sculpture on a museum pedestal, or of an architect's model of a future metropolis. There is all the difference in the world between such a gesture and a publicity photo of a hamburger fitting perfectly in a dish, as if it grew there from a seed and matured to deliciousness. The latter hides long labor within an image of ease, while the former "explodes" the cell-cultured hamburger into the tedium and care and patience of its creation. New Harvest did not mean to deflate hopes that cultured meat might one day move from such laborious handwork to an automated and industrial-scale process, but to show its audience where things stand. I have come to believe that such transparency is crucial for building public trust in a strange new world of food technologies. And I worry that projections for imminent cultured meat might jeopardize that trust if the companies cannot make good on them. Cultured meat may not survive another serious "trough of diminished expectations," as the hype cycle consultants like to put it. All this means that New Harvest is in a strategically difficult position. The technologies they wish to promote in a moderate spirit are being aggressively overpromoted by others. This sometimes makes New Harvest seem like anti-cheerleaders when they merely insist on keeping level heads, refusing to allow the emotional and ethical appeal of ending animal agriculture to muddy a clear accounting of the facts at hand. Not that those facts are easy to come by, given all the black boxes. Nor easy to interpret, when you're caught between those countervailing twined forces, doubt and hope, wishing they might achieve dialectical resolution in a third, more level, feeling. But doubt and hope rarely work that way. Usually one just beats the other. This has emerged as the journalistic and popular expectation at this juncture: either cultured meat will emerge soon or it won't, an either/or attitude that reflects the timetable of the start-ups.

Rumors were easy to catch as I walked around the large garden outside of Pioneer Works, where journalists, entrepreneurs, and other onlookers strolled around the rocks and trees and weighed in off the record. There were rumors that Just (formerly Hampton Creek), Mosa Meats, and Memphis Meats (or one of the three) have found a way around the growth medium problem and no longer need to use fetal bovine serum. Then again, there were also derisive sneers regarding the projected "2018" (Just) or "2019" (Mosa and Memphis) first-glimpse product release dates that had been suggested by the companies, not just because the technical bottlenecks still stood (as far as we knew), but also because of time-consuming regulatory hurdles. It isn't even clear which governmental body would regulate cultured meat. In the United States, the Department of Agriculture deals with meat, eggs, and poultry, while the Food and Drug Administration deals with a broad category called "biologics," including products made through tissue culture.[6] Those first products may be expensive pieces of meat in expensive restaurants, but they will be consumer-available just the same. Several of the companies have acknowledged that a release date for products available to a wide market would be more like 2021 than 2019.

When did cultured meat cease to be a story about an indefinite future, perhaps ten to twenty years out, and become a story whose narrative beats are counted in financial quarters? An easy answer is "When venture capital got interested." A fuller answer would note that the timetable required by investors resonates with the impatience of the animal protection activists, some of whom have grown tired of the slow progress made through protests, education, outreach, and lobbying. It is almost as powerful as the impatience of Willem van Eelen, whose daughter, Ira van Eelen, had just joined the board at Just. In September 2017, Just purchased the elder van Eelen's original patents and gained the younger van Eelen's help as well.[7] There are also rumors about the big food companies that have invested in cultured meat start-ups. Are they merely covering their bases against the possible viability of lab meat? Or do their strategists agree with one claim widespread in cultured meat circles, namely that climate change will soon increase the costs of conventional meat production to the point where it becomes, startlingly, less viable than cultured meat? They didn't send any representatives to make their will known at this conference.

The New Harvest 2017 stage was decorated with inflated pink and blue laboratory gloves grouped into spiky balls. On first seeing them, I mistook them for clusters of plastic udders in New Harvest's colors. As we took our seats for the afternoon session, the bio-artist Oron Catts (who, working with Ionat Zurr, produced the world's first cultured meat) offered me an interpretation I had never considered: cultured meat was born of failure. "This is the story of the failure of regenerative medicine," he said, meaning that regenerative medicine, another burgeoning frontier for tissue engineering, which made its way into the hype cycle years before cultured meat, has failed to meet the inflated expectations that formed around it. Catts's own interest in making bio-art was sparked by the same project that put regenerative medicine on the public's map: a naked mouse with something like a human ear growing on its back, unveiled by Charles Vacanti in 1995. The media made charismatic megafauna of the earmouse, and for more than two decades (or about the life spans of seven long-lived lab mice, laid end-to-end) Catts has been thinking of the public expectations that were raised: thousands of patients wait on organ transplant lists, and regenerative medicine could, in theory, save them all but shows no signs of doing so on those patients' schedules. Did medical tissue engineers gravitate toward cultured meat out of disillusionment with their former field, or out of a sense that cultured meat promised a greater ultimate "impact" on our world? Mark Post, Uma Valeti, and Vladimir Mironov are just three of the cultured meat scientists who started out by exploring the medical potentialities of stem cells. Several of the New Harvest fellows considered a career in medicine at one time or another. If the start-ups are as good as their word, there will be cultured meat products on the market before most of the fellows even finish their doctorates.

Over time, and under media pressure, the early claims made on behalf of regenerative medicine crystallized into a promise: by using a patient's stem cells, doctors could grow replacement organs, which could be transplanted into the patient with no risk of rejection.[8] Almost needless to say, the promise is not cashing out as quickly as some researchers had hoped or promised. The more damaging problem is that several scandals based on fraudulent scientific claims have loomed over both regenerative medicine and stem cell medicine more broadly.[9] Catts and I indulged in a moment of skepticism and I poured a little alcohol on the fire. What if a similar story could be told about biofuels? That form of "clean energy" enjoyed an investment boom in the late 2000s and then busted very publicly; cultured or "clean" meat may interest the same venture capitalists who were previously interested in biofuels and other "cleantech."[10] Not to mention the migration of talent from animal protection to the cultured meat start-ups and their supporting organizations. This movement suggests exhaustion with the project of changing hearts and minds and individual diets through the slow work of activism, and the hope that commercial activity might accomplish more, faster, and on a grander scale. Rather than changing personal dietary choices, the new strategy hopes to reduce or eliminate access to dietary choices that rely on cruelty to animals. Bruce Friedrich of GFI has gone on the record saying, "We are taking ethics off the table for consumers by making the sustainable and humane choice the default choice."[11] To say that cultured meat began with disappointments in other areas is not to say that it is doomed to the same fate as, say, regenerative medicine, which in any case has not yet accepted its doom and still moves forward. The point is that the characters and resources in this story have often been recruited from other tales, and they arrive trailing their own past histories. Failure is one way to tell this story, adaptation is another, and Catts's choice of the former is deliberate.

Cultured meat, I have sometimes mused when my mind turns to failure, might be a good idea at risk of being doomed by the specific form of investment and development in which it is presently caught up. The start-up model itself may jeopardize cultured meat, venture capitalists needing to see a return on their investment faster than laboratories can yield a viable product at scale. All this assumes that cultured meat is defined as something that not only mimics meat but is also made of animal cells, as opposed to the very sophisticated plant-based burgers coming to market as of 2017. It would follow that cultured meat needs a more patient form of capital, or that research should take place at the potentially slower pace of academic labs drawing on (one hopes) government funding. But even if this picture of a potentially compromised good idea turned out to be accurate, it would do little to describe cultured meat's relationship to capitalism.

Cultured meat stands to alleviate the suffering of food animals packed together at industrial scale and to reduce the massive environmental bootstamp of industrial animal agriculture, not to mention eliminating zoonotic diseases that proliferate in microbial feedlot ponds. But, as has been obvious since the start of our story, those problems are themselves symptomatic of deeper civilizational troubles. They derive from the advent of a global human population of unprecedented number, enjoying not only animal skeletal muscle daily or near daily, but also the other blessings of industrial civilization, down to the refrigerators in which meat is stored and the electricity those refrigerators run on, not to mention the gas mains leading to the stovetops where the meat is cooked. Meat is cheap, but nothing becomes inexpensive in a vacuum. Capitalism is not the only economic system that supported the multiplex process called modernization, nor has it held any historical monopoly on the overpromotion of meat—Soviet Russia, for example, had to import animal feed at one point, so great was the Soviets' desire to increase meat production and consumption. And yet the spread of the carno-centric Western diet has most often been a behavioral adjunct to the spread of free-market capitalism, perhaps most dramatically symbolized by the opening of McDonald's locations in Moscow and Beijing.[12]

Cheap meat is now part of a network of things that includes government subsidies and regulatory activities, the consolidation of agricultural activities by a few very large companies, a collection of tools for inducing animal bodies to produce more flesh than they would in any state of nature, and more. This network hums along in an economic system whose operating premise is always "more"—the ongoing growth of markets. We live in a functionally if not always ideologically cornucopian world in which Malthusians always seem like wild-bearded critics holding up warning signs on the sidelines, imposing unexpected and unwished-for neural burdens on people unused to hearing that babies and meat are anything but automatically good.

Although none of its architects would describe it in these terms, cultured meat promises to remedy one of the central problems of capitalist modernity itself. Call it cheapness, the ubiquitous cost-transference phenomenon of which cheap meat is just one example.[13] The cheapness of desirable things (food, fuel, clothing, labor, and so forth) can be hard to understand as a problem precisely because it has democratized access to material goods and helped to improve standards of living worldwide. Cheapness is the water in which we swim, so we don't notice it. But the "cheapening" of meat has meant much more than meat's democratization. It has meant shifting costs incurred in producing and distributing meat so that they are less visible to the consumer and less damaging to those who own the means of meat's production. Those costs are shifted to those who work menial jobs in the meat industry, and to the environment, and to food animals themselves in terms of their quality of brief life. Harm becomes invisible, a fact that animal protection activists have known for years. This is why some activists smuggle cameras into feedlots and slaughterhouses. Indeed, their work is to make harm to animals visible and ensure that humans are properly attuned to recognize it as harm. Complex production and supply chains also conceal human suffering, beyond the health burdens caused by meat-heavy diets. As one attendee at New Harvest 2017 pointed out to me, no one at the conference podium mentioned that, in the American South, African Americans are disproportionately hurt by water polluted by pig farming.[14] Cheapness could also be called damage that does not concern the powerful.

In a utopian future of cultured meat, tissue culture and engineering could make meat's cheapness unproblematic by greatly reducing meat's environmental and moral externalities. But a utopian future in which cultured meat scales and sells would also be one in which omnivorous civilization is maintained in something like its present shape, with greater prospects for sustainability than anyone living in 2017 can imagine. It is crucial to keep all this in the conditional. I cannot count the number of contingencies that would lie between technological success in cultured meat and such a utopian future. Cellular agriculture might be best described as a kind of market-driven utopia born out of the disasters that modernization has produced. In other words, if industrial animal agriculture is one of the forces making a civilization rooted in growth-oriented capitalism unsustainable, then through the business success of cellular agriculture, the market might just save growth from itself. In his "Fifty Years Hence," the same document in which he mentioned the dream of growing chicken parts in vitro, Churchill speculated in similar terms:

> If gigantic sources of power become available, food would be produced without recourse to sunlight. Vast cellars, in which artificial radiation is generated, may replace the cornfields and potato patches of the world. Parks and gardens will cover our pastures and ploughed fields. When the time comes, there will be plenty of room for the cities to spread themselves.[15]

Churchill's implication (void of fact as it may have been) was that a change of agricultural substrate would yield both more parks and more gardens: "plenty of room." Like Churchill's irradiated underground cellars, cellular agriculture seems to present a new kind of interior frontier, enabling fresh investment, profit, and growth at a moment when all frontiers are already developed and our existing agricultural land is under threat. As natural resource depletion and climate change increase the cost of animal agriculture, cells (so this story line runs) eventually become cheaper workers than whole-bodied animals. Cell lines assume an economic role once played by the cows, pigs, and chickens they come from. But, at least at the moment, cells are not cheaper workers than whole animals, and to assume that they one day will be constitutes a considerable gamble on that confused concept, technological progress.

When I stepped onto the New Harvest 2017 stage myself, I said none of this. I was there to give a short talk on the histories of meat and of tissue culture as they intertwine in cultured meat, placing special emphasis on the unpredictable and systemic character of changes in our foodways. My point was merely that it is impossible to see and narrate the full arc of a new food's emergence, spread, and ultimate effects from within the vanguard of its creation. Nor is it possible to see how cultured meat might change the way we see animal bodies or agriculture itself. In this spirit I made my little plea for contingency and irony.

I had the good luck that Catts took the stage immediately after me, to describe his work with Zurr at SymbioticA, "an artistic laboratory dedicated to the research, learning, critique and hands-on engagement with the life sciences."[16] Catts has been making art through tissue culture techniques for more than twenty years. Near the start of their work, Catts and Zurr collaborated with Joseph Vacanti (part of the "earmouse" project), and now Catts stood on New Harvest's stage, saying to his audience, "You are up there, and I am down here," indicating first a high point on a hype curve and then its opposite low point. It is often hard to tell how much of Catts's skepticism about laboratory-grown meat is dispositional, how much is ideological, and how much is just the earned skepticism of someone who watched a technology fail to emerge over a long period of observation.

Meanwhile, Catts makes what he and Zurr call their "semi-living" creations. Semi-living? The prefix means "half" in Latin, but in common English usage it carries the additional sense of an incomplete or partial process. Cultured meat scientists say much about the liveliness of the cells that divide and proliferate in their bioreactors, but not much about what it means for those cells that their life processes unfold in vitro. Catts and Zurr's compound term "semi-living" does different work. It emphasizes the distinction between life processes in vitro and in vivo, between what it means for integral organisms to live and what it means for pieces of them to persist and grow in glass or plastic under carefully controlled conditions. For what does it mean to say that cells "thrive" in vitro? Catts and Zurr scavenge raw biological materials for their work from other laboratories, taking tissue samples from recently killed laboratory animals and eking out a little postmortem life from the animals' cells. The very first steak they grew, prior to creating their 2003 frog steaks, was made from skeletal muscle cells taken from an unborn baby sheep.[17]

One of Catts and Zurr's most recent sculptures, made in collaboration with designer Robert Foster of Fink Design, was called *Stir Fly*. Displayed at Science Gallery Dublin, *Stir Fly* exploded unintentionally, harming no one. Before it self-vanquished, the artwork first challenged its viewers' assumptions regarding the nature of meat. The cells growing in the *Stir Fly* bioreactor came from insects, its growth medium included fetal bovine serum, and some twenty liters of that growth medium were suspended in a bag above the bioreactor, a fluid sword of Damocles. It can be useful to look at cultured meat with a tissue artist, perhaps particularly because of the public relations push to abandon all that is strange about cultured meat, including casting aside the older terminology of "in vitro" meat. Datar, speaking in a different context, has pointed out the regulatory value of such normalization efforts: "Most food regulation is about aligning new products with something that's already recognized as safe."[18] And normalization also supports the goal of eventual consumer acceptance. Catts and Zurr grow pieces of tissue not so unlike the earliest pieces of in vitro meat. In conversation, Catts talked to me about the similarity between tissue cultures grown for art and tissue grown as food and submitted that he and Zurr have found an aesthetic potential in biotechnology that the biotechnologists don't seem to want to acknowledge. Biotechnology might unwittingly perform one of art's functions and transform our way of seeing the world—more specifically, our way of seeing the meaning of biological life.

Philosophers will, as you might expect, debate the question of whether "meaning" is the sort of thing biological life possesses; here we should acknowledge the distinction between philosophical debate and the way we often experience the world. The small but eye-catching genre of bio-art has served double duty as a form of speculative futurism that asks not only what animal bodies might become, but also how living with new forms of life might change our perspectives. Bunnies can glow in the dark if they are (as a fertilized egg) implanted with jellyfish DNA for phosphorescence, but how would keeping such a pet change the way you see the world?[19]

Catts walked the audience through a slideshow of his and Zurr's past projects, from early experiments conducted at a time when art galleries were usually unwilling to display "wet" or "semi-living" art, up to the frog steaks grown for the project *Disembodied Cuisine* in 2003. Those steaks could easily have gotten the attention of Jason Matheny, who went on to found New Harvest in 2004, in a striking case of life imitating art (albeit to very different and more optimistic purposes). I don't know if Matheny had heard of the pair of wing-like pieces of pig bone that the Tissue Culture & Art Project (the name of Catts and Zurr's ongoing collaboration) grew during a residency at Harvard Medical School, a witty play on the feasibility of many biotechnology projects. The sculpture conveyed the message *When Pigs Fly* with minimal flapping. Projects like *Stir Fly,* or like the handful of *Semi-living Worry Dolls* that Catts and Zurr displayed at Science Gallery Dublin many years before *Stir Fly,* are not designed to fit classical aesthetic criteria for beauty.[20] If one purpose of Catts and Zurr's work is to encourage productive worry, another is to encourage feelings of care for surprising objects, such as the haphazardly formed worry dolls. What does it mean to extend care to "semi-living" things? In the case of the worry dolls, the idea was that the art would care back. The dolls, inspired by the worry dolls of Guatemalan folk art, would perform their traditional function of taking away the worries that are whispered to them. A microphone located outside the worry dolls' microgravity bioreactor carried museum visitors' personal concerns to the dolls' growing "ears." As Catts and Zurr explained, each doll had an identity defined by a specific worry. These ran alphabetically from A to H, beginning with "the worry from Absolute truths, and of the people who think they hold them," going on to "the worry from Biotechnology, and the forces that drive it" and "Capitalism, Corporations," and ending with "our fear of Hope."

SymbioticA's name resonates with a very different form of reciprocal care, namely symbiotic relations between organisms and cells of various kinds, which the microbiologist Lynn Margulis hypothesized as a driver of evolutionary change. According to Margulis's widely accepted view, the nucleated eukaryotic cells familiar to us emerged through bacteria absorbing each other and incorporating each other's structures into themselves, gaining such useful functions as the capacity for cellular respiration.[21] The resulting endosymbiosis merges genetic lineages laterally and constitutes an evolutionary force that Margulis suggested might rival the importance of natural selection and mutation in transforming species over time.[22] Reciprocal care, "SymbioticA" implies, may not only be an ethical stance toward life that humans may take. It might be an evolutionary process that made complex organisms, like us, possible. The *Worry Dolls* project, among other semi-living sculptures, shifts care into the domain of biotechnology. If many projects in applied biotechnology emphasize the control of nature, Catts and Zurr's emphasis on care suggests reciprocity instead.

Despite their indirect influence on the cultured meat movement, the central focus of Catts and Zurr's work is not biotechnology but our relationship with the concept of life itself, which Catts has described to me as "indeterminate" and currently characterized by instrumentalism. Biotechnology happens to be the current means by which life is made into an instrument. Humans may have bent life toolward when our ancestors domesticated the first wild animals and put them to work, but SymbioticA suggests that life is instrumentalized in a qualitatively different way when in vitro techniques allow the isolation of cells and tissues from the body. This is not dog breeding. Catts and Zurr never imply that life has a knowable albeit hidden essence that biotechnologies, whether genetic or somatic, somehow disrupt or corrupt. Theirs is not an art that celebrates a version of "pure" life undetermined by human will and technological means, or life as a veiled *mysterium tremendum*.[23] Nor is theirs an art that provides answers to the question of how we should or should not use biotechnology on animal bodies or on human ones. But they effectively reject one claim that subtends the cultured meat movement, namely that the life processes of cells are simply reducible to the movements of a mechanism, subject to control and optimization. They don't deny that life can seem that way. Rather, they ask what the analytic reduction costs us and what it buys us.

The flesh growing in Catts and Zurr's bioreactors is also an invitation to consider a different and more strictly philosophical question, one over which moral philosophers have taken conceptual pains: Are the realms of nature and human morality different realms, and how do humans live as simultaneous denizens of both? Is the former a realm of rules in which we are bound, much like all the other animals? Is the latter a realm in which our freedom distinguishes us from the rest of *Animalia,* as the creatures that can most fully name their aspirations? Some philosophers have insisted that we are totally free in our morality, while others argue that we are hemmed in by a transcendental version of the Good. Call it God or Nature, perhaps, or simply Reason, depending on the philosophical school you follow, and make the appropriate adjustments. The implication is that morality is not merely something we posit as we wish, either as individuals or by social convention or fiat, but that it is something we try to get right, at least at the level of aspiration. Thus did the philosopher Philippa Foot compare good actions to good roots, thinking of trees.

Goodness, Foot argued, against such influential twentieth-century British philosophers as G. E. Moore, was akin to a natural property.[24] As such it exists outside of human conventions. In Foot's essays of the late 1950s, she explained that moral philosophy had been preoccupied with a distinction between factual description and evaluation, the former being evidence based, subject to confirmation, and not at all subjective, whereas the latter is as subjective as a preference for coffee ice cream. Moral philosophers had converged on the view that moral statements were evaluative and subjective rather than factual, and Foot contested that view. She used the example of a knife, whose goodness as a knife presumes a specific functionality, to argue that "goodness" is not an arbitrary subjective claim about an object in question but recalls the essential purpose for which it was made or grown. Thus might we understand the goodness of eyes and lungs.[25] We could say many "evaluative" things about eyes and lungs and knives, possibly related to their aesthetic qualities (although it is hard to imagine privately held aesthetic views about hidden organs like lungs) but think of the "evaluative" judgments we might have about a gorgeously inlaid blade. In answering the question of whether this was a good knife, we might have to admit that it was more ornamental than useful. There are oyster knives, eel knives, boning knives, and so forth, each made to a purpose, and thus to a goodness. Roots, too.

Foot's virtue ethics (as it is called) might seem to rest on nature in a way biotechnology renders problematic. She writes that "the grounding of a moral argument is ultimately in facts about human life."[26] In the lab, facts seem to change and nature seems less a reliable undetermined foundation and more a narrow bridge, one that we must continually patch as we walk across it. What happens to the line between the realm of nature and the realm of freedom when we demonstrate that we are free to redesign small parts of nature, however laboriously or expensively? Does this mean that we should no longer use the natures of things as a way to think about their goodness? This is the philosophical problem Catts and Zurr ask us to consider.

But this philosophical problem does not float in a void. The encounter with the work of art supplies our problem's context. The work of Catts and Zurr's Tissue Culture & Art Project, and of SymbioticA, places human beings on the opposite side of the glass from a distinctly different form of "semi-living" life. This has two powerful effects. On one hand, we are confronted by sameness, for we are made of cells and tissues much like the "semi-living" sculptures, and the poignancy of those works lies in their conveyance of the mortal vulnerability of flesh, whether in vivo or in vitro. On the other hand, the "semi-living" is a grand "other" in the living sea of possible other beings against which we might define ourselves. We have integral bodies at the level of trunk and limbs and head, more or less, and we also have the experience of moral freedom. We perceive ourselves as sapient choice-makers in a fashion that most other creatures (much less tissue cultures) seem not to be. We created the CAFOs and the slaughterhouses and the bioreactors too, and we have the capacity (however little we exercise it) to reform animal agriculture in all manner of ways, or to reduce our consumption of animal products. To set the question in other terms, what does it say about the character of our moral freedom that we might bring the realm of nature within our freedom's compass?

New Harvest displayed bravery in inviting Catts to speak, not only because of his status as a cultured meat skeptic, but also because the work of Catts and Zurr challenges the idiom of imitation so central to the dominant discourses around cultured meat. In the rush to present cultured meat as a clean, familiar, and viable product, everything strange about it gets suppressed, often with a smile. Catts and Zurr's works invite us to think beyond culinary patterns that feature familiar cuts and parts; I think of that *gayal* again, and of the question of just how malleable our habits regarding meat really are. In his essay "Imitation of Nature," Hans Blumenberg supplied a program for understanding the Tissue Culture & Art Project some fifty years in advance.[27] In a compressed theory of modernity—modernity as we experience it—Blumenberg surmised that our disquiet about living in an increasingly artificial world, close by our creations, has to do with the distinction between the imitation of nature and acts of new invention. We once created things that we perceived as imitative of natural forces, beings, or processes, and we subsequently rebelled against nature by creating things for which the only model image is contained in the human mind. Blumenberg argued, through a discussion simultaneously Aristotelian and Biblical, that we are rebels ill at ease in the aftermath of our rebellion. This applies directly to growing meat in bioreactors, which begins with the probably correct assertion that only by copying conventional forms of meat could we win carnivorous appetites away from the feedlots, but which then encounters trouble. That trouble consists of sheer technical complexity, but not only that: the corollary to the technical difficulty of producing cuts of meat identical to, say, a haunch of *gayal,* is that producing less familiar and thus more "rebellious" forms of protein might be much easier. The future of cultured meat might be weirder than public relations campaigns can tolerate.

There was more. After a short break (more snacks, more conversations), a large panel discussion brought Illtud Dunsford, Mark Post, and a New Zealand farmer named Richard Fowler together with David Kay, a representative of Memphis Meats, to discuss the relationship between cultured meat and conventional farming. Danielle Gould, founder of an organization called Food+Tech Connect, which adopts a tech-industry-style "hackathon" approach to food challenges, moderated. We expect a certain amount of tension from such conversations, and I was reminded that one survey of potential consumer attitudes toward cultured meat, in the United Kingdom, elicited the response that it seemed like the sad "end of a system," the end of agricultural relations between humans and animals.[28] The end of a moral ecology of production, you might say. Post has gone on record saying that he does not believe that cultured meat and conventional animal agriculture could coexist; the new, he suggests, would end the old. But the tone of the conversation was largely friendly. Kay was eager to say that Memphis Meats only wishes to "disrupt" industrial animal agriculture. Dunsford pointed out that without farmers, we lose the stewards of the land. The microphone kept being passed and a conciliatory theme emerged, namely that farmers and the cellular agriculture industry will benefit from working together. Despite his belief that conventional and cultured meat industries would be basically incompatible, Post suggested that many raw vegetable ingredients for a cellular agriculture industry will come from conventional allies; his is not a nineteenth- or early twentieth-century utopian vision of food production fully cut off from the fields.

Then a sharp crack shook the room, and the power went off for a moment. I heard the usual surprised chortles of an audience for whom untroubled access to electricity, perhaps especially for their phones, is an assumption of everyday life. The power came back on and everything went back to normal. Dunsford said more about stewardship. Post reflected that artisanal-scale conventional meat production might proceed in a world of cultured meat, though it would be a niche activity, heavily subsidized, and not really a functional sector of the economy. (This strikes me as a very European point; in many parts of the European Union, artisanal food production receives government subsidies, particularly in countries like Italy and France that value agricultural heritage.) They passed around the term "clean meat," opining on its value, and I observed how divisive it has become. For many, calling cultured meat "clean" implies that conventional meat is impure, making it a form of indirect insult. It suggests a moral judgment about consumers' current practices. Kay noted that Memphis Meats uses the term, and I recalled that Memphis Meats is closely allied to the GFI, which champions "clean meat." Post claimed terminological agnosticism, but he joked that "clean meat" doesn't translate well into Dutch. There was a wonderful question from the audience: "It is 2050, and cellular agriculture is in effect. Describe our relationship with the natural world." Responses ranged from vastly diminished global herds of food animals, to carefully maintained small herds designed to preserve genetic diversity (after all, there is a critical difference between cellular copying and sexual reproduction), to a vestigial animal agriculture industry a bit like the limited-term coexistence of horse-drawn carriages and automobiles in the early twentieth century.

As Yuki Hanyu took the stage to discuss his work at Shojin Meats, I reflected on what I had seen. Cultured meat is still emerging as I finish writing this book, observing the events around me with wild surmise. Promises are made on behalf of cultured meat, and as chroniclers like myself measure out doubt and hope, basic research moves forward at its appropriately slow pace. The general tropism is toward success, and I realize I do want them to succeed—I want Isha Datar and New Harvest and Mark Post and their colleagues to get something resembling the futures they want. But I cannot know if they will. And if all this talk of lab-grown meat turns out to remain in the realm of science fiction, then I hope it is not the escapist kind. Let it instead perform the classic function of the best science fiction by serving as a mirror on the present. If I have been an ironical guide, it is not to mock the sincerity of those who wish to solve problems. I have been taking lessons in sincerity in the field, and I have learned enough to think that sincerity and irony are far from incompatible modes of expression. To dismiss the hope for new technologies outright would be easy and cheap, a contemporary form of cultural pessimism that leads us to ignore forces that can change our world. I admit my skepticism and my reservations about whether a new mode of food production is the right way to resolve a set of problems that seem to me not merely technical, but social and political.

On the morning of the second day of the conference, I thank Jordan for the diagram and head back to Brooklyn, uncertainty and reciprocity on my mind.

As of mid-2018 we can start to get a sense of the strange object, made of cultured animal muscle cells and hopefully of flavor-bearing fat as well, that is trying to emerge from the fog. Its full emergence is not yet guaranteed, but trumpets blare to welcome it. The image of the sacrificial *gayal* in Lévi-Strauss, though, reminds me that cultured meat in its present form is anything but meat as an image of reciprocal and communal relationships—an image of the genesis of political life. Nor, of course, is the vast majority of the conventional meat I might buy in a store. For many of its advocates, cultured meat resolves the issue of our obligations to other creatures (if we kill and eat them, what do we owe them beforehand?), but it raises other issues at the same time, by implying that a world we have made, an even more artificial world than the one we already inhabit, is sufficient. Not merely sufficient for our animal needs but for our human aspirations, in all our understandable confusion and debate over what to name them.

Names, incidentally, were much in the cultured meat news not long after the New Harvest 2017 conference. On February 2, 2018, the United States Cattlemen's Association (USCA) issued a petition asking the Department of Agriculture to define two words, "meat" and "beef." Seemingly drafted by the USCA's lawyers, the document asks for definitions of "meat" and "beef" that explicitly protect the interests of cattlewomen and cattlemen while prohibiting a variety of products, including cultured meat, from being labeled as "meat" or "beef." The document at first seems to makes its case in a straightforward and detailed fashion. Its complexity shows up only after repeated readings, and its most significant complexity is this: the USCA acknowledges that "currently, there is no definition of what constitutes a 'beef' or 'meat' product." Naturally, the authors of the document—entitled

PETITION FOR THE IMPOSITION OF BEEF AND MEAT LABELING
REQUIREMENTS: TO EXCLUDE PRODUCTS NOT DERIVED DIRECTLY
FROM ANIMALS RAISED AND SLAUGHTERED FROM THE DEFINITION
OF "BEEF" AND "MEAT"

—do not appear to be motivated by ontological distress about definitions. As the petition acknowledges, it was designed to protect an existing industry from new arrivals. The new arrivals named in the petition are both plant-based and lab-grown meats, though meats of insect origin also receive mention. The authors make their case for the existence of a real and present danger to cattlemen, citing specific start-ups that are trying to make ersatz "meat," as well as major investors in those companies, including meat industry giants Tyson and Cargill. Toward this end the document's authors describe a new factory built by the plant-based burger company Impossible Foods, located in Oakland, which is projected to produce as much as twelve million tons of plant-based meat per year. We are to understand that "traditional" "beef" and "meat" are under threat in both semantic and financial terms.

The word "traditional" and the associated phrase "in the traditional manner" are repeated unto a mantra in the petition. They seem to form a bulwark against the legitimacy of cultured meat as meat or beef. The authors don't define the "traditional manner," but it seems to refer to animals who are bred and raised and fattened and slaughtered, disqualifying any cow cells grown in vitro from being beef or meat, however closely they may resemble meat grown in vivo. "The traditional manner" thus resists the rhetorical effort made by many proponents of cultured meat to demonstrate that animals and bioreactors, in vivo and in vitro, are basically equivalent. But the reliance on an undefined term, "traditional," gives the petition a circular character reminiscent of the associations carried by the term "meat" itself. The meaning of "meat" has changed throughout human history, and yet we have come to use it to indicate solidity of meaning, seriousness of content ("the meat of the matter"), or even a certain predictability of character ("a meat-and-potatoes man"). "Meat" is a mostly empty signifier that always appears full. Nevertheless, the petition gets passed around the cultured meat movement with a sense of excitement. Such an early "warning shot" from one corner of the meat industry could mean that the cattlemen are scared, not only by the possibility of cultured meat but also by Impossible Foods, Beyond Meat, and other companies that have recently created convincing plant-based foods that mimic the texture and taste of hamburger.

My favorite idea to come out of the world of cultured meat is the "pig in the backyard." [29] I say "favorite" not because this scenario seems likely to materialize but because it speaks most directly to my own imagination. In a city, a neighborhood contains a yard, and in that yard there is a pig, and that pig is relatively happy. It receives visitors every day, including local children who bring it odds and ends to eat from their family kitchens. These children may have played with the pig when it was small. Each week a small and harmless biopsy of cells is taken from the pig and turned into cultured pork, perhaps hundreds of pounds of it. This becomes the community's meat. The pig lives out a natural porcine span, and I assume it enjoys the company of other pigs from time to time. This fantasy comes to us from Dutch bioethicists, and it is based on a very real project in which Dutch neighborhoods raised pigs and then debated the question of their eventual slaughter. The fact that the pig lives in a city is important, for the city is the ancient topos of utopian thought.

The "pig in the backyard" might also be described as the recurrence of an image from late medieval Europe that has been recorded in literature and art history. This is the pig in the land of Cockaigne. Cockaigne, the "Big Rock Candy Mountain" of its time, was a fantasy for starving peasants across Europe. It was filled with foods of a magnificence that only the starving can imagine. In some depictions, you reached this land by eating through a wall of porridge, on the other side of which all manner of things to eat and drink came up from the ground and flowed in streams. Pigs walked around with forks sticking out of backs that were already roasted and sliced. Cockaigne is an image of appetites fulfilled, and cultured meat is Cockaigne's cornucopian echo. The great difference is that Cockaigne was an inversion of the experience of the peasants who imagined it: a land where sloth became a virtue rather than a vice, food and sex were easily had, and no one ever had to work. In Cockaigne, delicious birds would fly into our mouths, already cooked. Animals would want to be eaten. By gratifying the body's appetites rather than rewarding the performance of moral virtue, Cockaigne inverted heaven.

The "pig in the backyard" does not fully eliminate pigs, with their cleverness and their shit, from the getting of pork. It combines intimacy, community, and an encounter with two kinds of difference: the familiar but largely forgotten difference carried by the gaze between human animal and nonhuman animal, and the weirder difference of an animal's body extended by tissue culture techniques. Because that is literally what culturing animal cells does, extending the body both in time and space, creating a novel form of relation between an original, still living animal and its flesh that becomes meat. The "pig in the backyard" tries to please both hippies and techno-utopians at once, and this is part of the charm of this vision of *rus in urbe*.

But this doubled encounter with difference also promises (that word again!) to work on the moral imagination. The materials for this work are, first, the intact living body of another being, which appears to have something like a telos of its own beyond providing for our sustenance; and, second, a new set of possibilities for what meat can become in the twenty-first century. The "pig in the backyard" is only a scenario. Its outcomes are uncertain. It is not obvious that the neighborhood will want to eat the flesh, even the extended and "harmless" flesh, of a being they know well, but the history of slaughter and carnivory on farms suggests that they very well might. The "pig in the backyard" is an experiment in ethical futures. The pig points her snout at us and asks what kinds of persons we might become.

EIGHTEEN

Epimetheus

A spear transfixes a chicken nugget. One year after the New Harvest conference in New York, an advertisement appears in my Internet browser. "Lay down your spears" reads the slogan. The ad is from a company that has claimed it will put cultured meat on the market by the end of 2018, but it is now October 17, 2018, and the year is running down. At this point, I am less interested in such promises and more drawn to the semiotics of the ad: ancient weapon meets industrial food product, implying a continuity between our ancestral past as hunters, our present-day status as eaters of cheap industrial meat, and our possible future as consumers of a lab-grown replacement. It's five years after Mark Post's hamburger demonstration and some things haven't changed a bit, including the habit of collapsing time in the promotion of cultured meat, as if we were still hunter-gatherers, as if a chicken nugget had anything to do with an auroch or a deer. In the ad, the phallic Neolithic pierces the industrial modern, as if an appetite for meat were hardwired into human nature. A message from the company's CEO accompanies the image: "400,000 years ago, meat became part of the human diet, and throughout time, human beings have needed to kill the animal to enjoy their meat. First, with spears. Then, with industrial machines. Get ready for that paradigm to change."[1]

The ad collapses millennia in order to catch the viewer's eye, implying an arc of technological progress from the spear to the bioreactor. I think, and not for the first time, of the femur bone in *2001: A Space Odyssey,* used as a club, that spins through the air and becomes a spaceship in the next shot. Call it the product of a ballistic imagination.[2] But the ad also prompts a realization. For an urban modern like myself, it is harder to imagine a world in which all meat is obtained by hunting than it is to imagine meat growing in gleaming bioreactors. Like many people, I spend less time with animals than I do with machines. Promethean feats of technological progress reach the newspapers each week, but the work of hindsight, of reconstructing how our distant ancestors lived, might be harder than the work of dreaming about the future. Hindsight, however, gets a bad rap when compared to foresight. The term "hindsight" usually implies not historical understanding, but a state of regret. The name of the titan Prometheus means "forethought," whereas his brother Epimetheus's name means "hindsight" or "late counsel," and Epimetheus is often depicted as his brother's "shadow" or lesser image, the fool who eventually marries Pandora, bearer of the infamous box. In the *Protagoras,* Plato explains that Epimetheus and Prometheus were tasked with modifying and enhancing the creatures the gods had made out of a mixture of fire and clay, so that those creatures might have the necessities for survival. Epimetheus immediately began handing out scales, flippers, wings, claws, and so on to all the animals, and besides that he made prey species more numerous than predators.[3] But when he came to humans he had nothing left to give, leaving a naked species to fend for itself, and necessitating his brother Prometheus's theft of fire. Even today, there is less money to be made in the hindsight business.

The story of Epimetheus and Prometheus is on my mind as more news rolls in, news of climate change and of meat's implication in that process. Disastrous forms of climate change are likely to arrive by 2040, suggests a recent report from the United Nations, if we do not turn "the world economy on a dime," transforming our means of production and consumption "at a speed and scale that has 'no documented historical precedent.'"[4] Meanwhile, a study with the modest title "Reducing Food's Environmental Impacts through Producers and Consumers" has confirmed, through an examination of nearly forty thousand farms worldwide, that animal agriculture does the preponderance of agriculture's environmental harm, despite animal products providing a smaller percentage of our calories than plants.[5] The political philosopher Leo Strauss called Epimetheus "the being in whom thought follows production,"[6] but as I consult these reports I have to wonder if Prometheus might have failed to display forethought in stealing fire from Olympus and giving it to humanity. After all, in the industrial order eventually brought about by Prometheus's gift of civilizing fire, long-term thinking certainly seems to trail behind the mode of food production in which we've become enmeshed.

The science fiction writer William Gibson has said that "novels set in imaginary futures are necessarily about the moment in which they are written. As soon as a work is complete, it will begin to acquire a patina of anachronism."[7] He has also extended this idea beyond science fiction. "Every vision of the future," he has argued, "begins to obsolesce upon conception."[8] I have tried to write this book with Gibson's "obsolescence upon conception" in mind. This notion surely holds true for all the visions of the future of cultured meat I have documented. All the fieldwork I have done will come to seem antiquated in the run of time. From the standpoint of a future reader, cultured meat will have either succeeded or failed. Flesh grown in bioreactors will disappear from the scene or it will become normal, part of the unexceptional everyday.

While I have examined cultured meat as a very real technological project, following Gibson I have also treated it as a piece of science fiction, a mirror on the present.[9] Call this book a biotechnological nature walk, an assemblage of detours through the history of the future of food, a collection of meditations on meat, attentive not only to the ideas of scientists and engineers but also to the way they serve as catalysts for philosophical, anthropological, and historical inquiry. Not to set up manifestos for the future, but so that we might better know ourselves today. This is more than an effort to "inoculate" my work against inevitable obsolescence. The moral claims embedded in cultured meat remind us that conventional meat already incorporates moral claims, though we may not notice them because meat is so familiar. To butcher a pig is a moral claim. To price its meat is also a moral claim, because it establishes social limits on who has access to this meat. And, by extension, to produce a carbon footprint first by raising the pig, then by shipping its meat from the point of production to a meat counter in a faraway city, constitutes a moral claim regarding the environment. That these claims are cloaked under the name of everyday life does nothing to alter their moral character. Morality incorporates not only our active choices, but also our conscious reconciliation to, or unconscious acceptance of, cultural and social norms. We live our lives as dwellers within an implicit moral architecture.

As of this writing, cultured meat is still an emerging technology, neither foreclosed nor guaranteed, and its moral dimension is still apparent, explicit, allowing it to reflect the moral claims embedded in cheap industrial meat. That makes this a particularly propitious time for debate about the character of our food system. Foresight and hindsight are both ready at hand. Technology contours our moral options, but we retain some agency to shape the technological systems in which we live, and food systems are notably technological constructs. But they are also political constructs, and if I were so bold as to distill this book's eighteen chapters into a set of "theses on the future of cultured meat," this point would loom large among them. Another thesis would be that cultured meat springs not only from the imagination of medical researchers, but also from a form of imaginative arbitrage. Our inability to think seriously about changing our food system through collective action and political will encourages us to turn to technology and the market for solutions. An entire industry may spring up out of an impoverished sense of possibility. Or, to put the point more kindly, one form of the imagination flourishes because another form of the imagination cannot.

Cultured meat illuminates new moral options for us, a third thesis might begin, but the core of this thesis would be that the very idea of "new moral options" implies that morality is subject to change over time. It is not absolute. The imagined cultured meat scenario called the "pig in the backyard," in which urban neighborhoods communally raise a pig and eat the meat made from biopsies of its muscle cells, implies that we might come to regard a porcine individual—whose relatives we previously killed and ate—as living within our circle of moral concern. The story of cultured meat thus far is not only a story about animal suffering, environmental protection, and sustainable protein. It is also a story about the mutability of moral concern, and the role of technology in changing morality's horizon. This further suggests that we are not only responsive to our sense of morality, but in some senses responsible for its contents. Whether we define our morality in terms of the consequences of our actions or via laws, or by reference to a sense of virtue, there is no such thing as "moral progress" outside of our shifting, and necessarily collective, definitions of that progress. This thought makes me anxious about the prospect of new technologies that arrive without public debate.

I have written this book in a state of what you might call "future fatigue," a cousin to what Alvin and Heidi Toffler once called "future shock."[10] I have absorbed a lot of promises. And I have come to realize that the cultured meat movement has been launched not during a time of general optimism about the future, but in one of worry and pessimism. News rolls in of rising sea levels, of human civilization having wiped out some 60 percent of wild animals since 1970, of a great plastic garbage patch floating in the Pacific, of toxic pools resulting from the production of smartphones, including the one that served me as a research tool in my pursuit of cultured meat, a little portal between cyberspace and meatspace. In the face of all this, you might think of cultured meat as an endeavor to re-enchant the future, to make it seem possible again, beginning by repairing the relationship between humans and other beasts, so that their cells and not their bodies grace our plates. Imagine that we have paid a visit to our neighborhood "pig in the backyard." Not just to say thank you for the roast pork, but also to share an apple with a fellow creature, to watch it root around its little parcel of land, and to remember that the uncompleted project of becoming what we might be starts with questions.

NOTES

CHAPTER ONE. CYBERSPACE/MEATSPACE

1. Promises, especially promises made on behalf of new forms of biotechnology, are a major subject of this book, and especially of chapter 2. On promises, see Mike Fortun, *Promising Genomics: Iceland and deCODE Genetics in a World of Specula-tion* (Berkeley: University of California Press, 2008), and Fortun, "For an Ethics of Promising, or: A Few Kind Words about James Watson," *New Genetics and Society* 24 (2005): 157–174.

2. See Walter Benjamin, "Paris, Capital of the Nineteenth Century," in *Reflections: Essays, Aphorisms, Autobiographical Writings* (New York: Harcourt Brace Jovanovich, 1978), 151.

3. The definitive history of tissue culture is Hannah Landecker's *Culturing Life: How Cells Became Technologies* (Cambridge, MA: Harvard University Press, 2007).

4. As of this writing, the terminology used to describe lab-grown meat is still changing. I have adopted the term "cultured meat" for several reasons, including its built-in reference to tissue culture techniques, but most importantly because it functions as a time stamp for the period in which I conducted my research and wrote this book, which we might call the "cultured meat period."

5. See Anna Tsing, "How to Make Resources in Order to Destroy Them (and Then Save Them?) on the Salvage Frontier," in Daniel Rosenberg and Susan Harding (eds.), *Histories of the Future* (Durham, NC: Duke University Press, 2005).

6. See Raj Patel and Jason W. Moore, *A History of the World in Seven Cheap Things: A Guide to Capitalism, Nature, and the Future of the Planet* (Oakland: University of California Press, 2018).

7. On the ubiquity of advertising on the Internet under late capitalism, see Jonathan Crary, *24/7* (New York: Verso, 2013).

8. See Matt Novak, "24 Countries Where the Money Contains Meat," Giz-modo.com, posted and viewed November 30, 2016.

9. See Richard Wrangham, *Catching Fire: How Cooking Made Us Human* (New York: Basic Books, 2010).

10. For a longer discussion of arguments linking meat to human evolution, see chapter 2. For a view contrary to Wrangham's, see Alianda M. Cornélio et al., "Human Brain Expansion during Evolution Is Independent of Fire Control and Cooking," *Frontiers in Neuroscience* 10 (2016): 167.

11. On sociobiology and its critics, see chapter 2. Using primitivism as a foil for futurism is, incidentally, old news. In 1968, the literary critic and media theorist Marshall McLuhan created a fashion spread in *Harper's Bazaar* in which African tribesmen carrying spears and female European models appeared side-by-side; the year before, in an essay in *Look* magazine, McLuhan had described "the student of the future" as "an explorer, a researcher, a huntsman who ranges through the new educational world of electric circuitry and heightened human interaction just as the tribal huntsman ranged the wilds."

12. But see Matt Cartmill's anthropological and cultural-historical study of hunting, *A View to a Death in the Morning* (Cambridge, MA: Harvard University Press, 1993), for an analysis of the notion of an arc linking our ancestors' love of meat—and more specifically their love of hunting—to modern technology. As Cartmill points out, in Stanley Kubrick's 1968 film *2001: A Space Odyssey,* an evolutionary and historical arc is rendered in a single image, as the zebra femur that one australopithecine uses to kill another is transformed, through the use of montage, into a spacecraft. See Cartmill, 14.

13. Orville Schell wrote an important and still relevant early survey of the problems of excessive antibiotic use in the U.S. meat industry. See Schell, *Modern Meat: Antibiotics, Hormones, and the Pharmaceutical Farm* (New York: Vintage, 1978). Recent research conducted near the massive feedlots of Texas suggests that both antibiotics and antibiotic-resistant bacteria may be transmitted aerially from those feedlots, a troubling development. On the cloud of controversy surrounding this research, including significant opposition from the cattle industry, see Eva Hershaw, "When the Dust Settles," *Texas Monthly,* September 2016. On the history of subtherapeutic doses of antibiotics used to encourage the growth of chickens, see Maryn McKenna, *Big Chicken: The Improbable Story of How Antibiotics Created Modern Farming and Changed the Way the World Eats* (Washington, DC: National Geographic Books, 2017).

14. See J. E. Hollenbeck, "Interaction of the Role of Concentrated Animal Feeding Operations (CAFOs) in Emerging Infectious Diseases (EIDS)," *Infection, Genetics and Evolution* 38 (2016): 44–46.

15. See Hanna L. Tuomisto and M. Joost Teixeira de Mattos, "Environmental Impacts of Cultured Meat Production," *Environmental Science & Technology* 45 (2011): 6117–6123. For a more recent assessment, see Carolyn S. Mattick, Amy E. Landis, Braden R. Allenby, and Nicholas J. Genovese, "Anticipatory Life Cycle Analysis of In Vitro Biomass Cultivation for Cultured Meat Production in the United States," *Environmental Science & Technology* 49 (2015): 11941–11949. But see also Sergiy Smetana, Alexander Mathys, Achim Knoch, and Volker Heinz, "Meat Alternatives: Life Cycle Assessment of Most Known Meat Substitutes," *International Journal of Life Cycle Assessment* 20 (2015): 1254–1267, in which the authors

find that for many species whose cells are grown, culturing meat would actually be more environmentally damaging than raising animals for slaughter, given the high energy requirements of production.

16. Wrangham has a substantial interest in aggression among primates, and especially among male primates. See his *Demonic Males: Apes and the Origin of Human Violence,* coauthored with Dale Peterson (New York: Houghton Mifflin, 1996).

17. See Leo Marx, *The Machine in the Garden: Technology and the Pastoral Ideal in America* (Oxford, UK: Oxford University Press, 1964).

18. See Fredrik Pohl and Cyril M. Kornbluth, *The Space Merchants* (New York: Ballantine, 1953). The work was first released in serialized form in 1952.

19. On the stem cell as a figure of potentiality, see Karen-Sue Taussig, Klaus Hoeyer, and Stefan Helmreich, "The Anthropology of Potentiality in Biomedicine," *Current Anthropology* 54, Supplement 7 (2013).

20. I borrow this phrasing from Stefan Helmreich; see Helmreich, "Potential Energy and the Body Electric: Cardiac Waves, Brain Waves, and the Making of Quantities into Qualities," *Current Anthropology* 54, Supplement 7 (2013).

21. "To breed an animal with the right to make promises—is not this the paradoxical task that nature has set itself in the case of man? Is it not the real problem regarding man?" Friedrich Nietzsche, *On the Genealogy of Morals,* trans. Walter Kaufmann (New York: Vintage Books, 1969), 57. For an extended reading of this fragment from Nietzsche, see chapter 3.

22. In the Netherlands, a label called "Beter Leven" is used, with a star rating of one to three, to indicate the degree to which a product's creation minimized animal suffering.

23. See Josh Schonwald, *The Taste of Tomorrow: Dispatches from the Future of Food* (New York: HarperCollins, 2012).

24. Food science expert Harold McGee defines the Maillard reaction, which produces many of the flavors characteristic of cooked meats, as follows: "The sequence begins with the reaction of a carbohydrate molecule ... and an amino acid.... An unstable intermediate structure is formed, and this then undergoes further changes, producing hundreds of different by-products." See McGee, *On Food and Cooking: The Science and Lore of the Kitchen* (New York: Scribner, 1984), 778.

25. I borrow this observation and phrasing from Steven Shapin; see his essay "Invisible Science," *The Hedgehog Review* 18 (3) (2016).

26. See chapter 2 for a discussion of meat's physical qualities.

27. Alexis C. Madrigal, "When Will We Eat Hamburgers Grown in Test-Tubes?" *The Atlantic,* August 6, 2013. Helpfully, Madrigal continued to update his list of predictions for several years following Post's hamburger demonstration; as of this book's writing, the most recent predictions include Post's continued but notably shorter predicted time frame for a consumer product becoming available, and those made by his competitors, including Uma Valeti of the San Francisco–based company Memphis Meats and Josh Tetrick of Just (formerly Hampton Creek). See a chart of

Madrigal's predictions at www.theatlantic.com/technology/archive/2013/08/chart-when-will-we-eat-hamburgers-grown-in-test-tubes/278405/, accessed April 25, 2017, and see his sources, listed at https://docs.google.com/spreadsheets/d/1yOT10HJwGVc9Ngkt2ar58Cp5W6CyeAWilPokf1lp_4Q/edit.

28. Here I paraphrase the anthropologist Marshall Sahlins, who once spoke of discovering "the lineaments of the larger society in the concepts of its biology." See Sahlins, *The Use and Abuse of Biology: An Anthropological Critique of Sociobiology* (Ann Arbor: University of Michigan Press, 1976), discussed in greater detail in chapter 2. On the intellectual history of biotechnology from the nineteenth to the twentieth century, see Philip J. Pauly's important *Controlling Life: Jacques Loeb and the Engineering Ideal in Biology* (Berkeley: University of California Press, 1987).

29. See, for example, Christina Agapakis, "Steak of the Art: The Fatal Flaws of In Vitro Meat," *Discover,* April 24, 2012, discussed in greater detail in chapter 5.

30. On the history of food futurism, see Warren Belasco, *Meals to Come: A History of the Future of Food* (Berkeley: University of California Press, 2006). I refer to Belasco's book, the only one dedicated to its topic, throughout this book, and especially in chapter 8.

31. See Fortun, "For an Ethics of Promising."

32. On the bricolage metaphor, coined by Claude Lévi-Strauss and since then traded upon in many fields within the human sciences, the humanities, and the natural sciences, see Christopher Johnson, "Bricoleur and Bricolage: From Metaphor to Universal Concept," *Paragraph* 35 (2012): 355–372.

CHAPTER TWO. MEAT

1. For one history of the polio virus, see David M. Oshinsky, *Polio: An American Story* (Oxford, UK: Oxford University Press, 2006); and see Hannah Landecker, *Culturing Life: How Cells Became Technologies* (Cambridge, MA: Harvard University Press, 2007), ch. 3: "Mass Reproduction."

2. See Vaclav Smil, "Eating Meat: Evolution, Patterns, and Consequences," *Population and Development Review* 28 (2002): 599–639; see 618.

3. See Henning Steinfeld et al., "Livestock's Long Shadow," FAO, 2006, www.fao.org/docrep/010/a0701e/a0701e00.HTM.

4. John Berger, "Why Look at Animals?" in *About Looking* (New York: Vintage, 1991).

5. See, for example, Hanna Glasse's 1763 edition of her *Art of Cookery, Made Plain and Easy,* 7th ed. (London, 1763), 370.

6. During the medieval period, Europeans of all social classes ate swans. It was only later that swans became associated with the feasts of the upper classes. By the twentieth century, swans had vanished from almost all European tables. In England, the royal family still enjoys special rights over swans, first formalized in the 1482 Act of Swans, although the royal family has periodically granted other institutions the

right to own and consume swans. St. John's College, Cambridge, for example, has such a right, and has been known to exercise it, serving swans at formal dinners.

7. See Charles Huntington Whitman, "Old English Mammal Names," *The Journal of English and Germanic Philology* 6 (1907): 649–656.

8. On cows as property in ancient Greece, see Jeremy McInerney, *The Cattle of the Sun: Cows and Culture in the World of the Ancient Greeks* (Princeton, NJ: Princeton University Press, 2010).

9. See, for example, Jillian R. Cavanaugh, "Making Salami, Producing Bergamo: The Transformation of Value," *Ethnos* 72 (2007): 149–172.

10. On the symbolism of British beef, see Ben Rodgers, *Beef and Liberty: Roast Beef, John Bull and the English Nation* (London: Vintage, 2004). On the history of the hamburger, see Josh Ozersky, *The Hamburger* (New Haven, CT: Yale University Press, 2008). See also James L. Watson (ed.), *Golden Arches East: McDonald's in East Asia* (Stanford, CA: Stanford University Press, 1997).

11. The architectural historian Reyner Banham writes of the hamburger and its appropriateness in Los Angeles, a city whose very foundation is movement: "The purely functional hamburger, as delivered across the counter of say, the Gipsy Wagon on the UCLA campus, the Surf-boarder at Hermosa Beach or any McDonald's or Jack-in-the-Box outlet anywhere, is a pretty well-balanced meal that he who runs (surfs, drives, studies) can eat with one hand; not only the ground beef but all the sauce, cheese, shredded lettuce, and other garnishes are firmly gripped between the two halves of the bun." Banham, *Los Angeles: The Architecture of Four Ecologies* (Harmondsworth, UK: Penguin Books, 1971), 111.

12. On animal breeding in Victorian Britain, see Harriet Ritvo, *The Animal Estate: The English and Other Creatures in the Victorian Age* (Cambridge, MA: Harvard University Press, 1987), ch. 2: "Barons of Beef."

13. For one examination of how modernization has affected Chinese carnivory, see James L. Watson, "Meat: A Cultural Biography in (South) China," in Jakob A. Klein and Anne Murcott (eds.), *Food Consumption in Global Perspective: Essays in the Anthropology of Food in Honour of Jack Goody* (Basingstoke, UK: Palgrave Mac-Millan, 2014).

14. See Loren Cordain, S. Boyd Eaton, Anthony Sebastian, Neil Mann, Staffan Lindeberg, Bruce A. Watkins, James H. O'Keefe, and Janette Brand-Miller, "Origins and Evolution of the Western Diet: Health Implications for the 21st Century," *American Journal of Clinical Nutrition* 81 (2005): 341–354.

15. Oron Catts and Ionat Zurr, "Ingestion/Disembodied Cuisine," *Cabinet* no. 16 (winter 2004/5). See also Catts and Zurr, "Disembodied Livestock: The Promise of a Semi-living Utopia," *Parallax* 19 (2013): 101–113.

16. See Harold McGee, *On Food and Cooking: The Science and Lore of the Kitchen* (New York: Scribner, 1984), 121–137.

17. Ibid., 129.

18. See, for example, Jacob P. Mertens et al., "Engineering Muscle Constructs for the Creation of Functional Engineered Musculoskeletal Tissue," *Regenerative Medicine* 9 (2014): 89–100.

19. See Carol Adams, *The Sexual Politics of Meat: A Feminist-Vegetarian Critical Theory* (New York: Continuum, 1990). See also Nick Fiddes, *Meat: A Natural Symbol* (London: Routledge, 1991).

20. See Fiddes, *Meat*.

21. On meat as food for heroes, see, for example, Egbert J. Bakker, *The Meaning of Meat and the Structure of the Odyssey* (Cambridge, UK: Cambridge University Press, 2013).

22. See Josh Berson, "Meat," Remedia Network, July 27, 2015, https://remedia network.net/2015/07/27/meat/, accessed March 28, 2017. See also Berson, *The Meat Question: Animals, Humans, and the Deep History of Food* (Cambridge, MA: MIT Press, forthcoming).

23. For example, see C.L. Delgado, "Rising Consumption of Meat and Milk in Developing Countries Has Created a New Food Revolution," *Journal of Nutrition* 133(11), Supplement 2 (2002). See also Josef Schmidhuber and Prakesh Shetty, "The Nutrition Transition to 2030: Why Developing Countries Are Likely to Bear the Major Burden," FAO, 2005, www.fao.org/fileadmin/templates/esa/Global_persepctives/Long_term_papers/JSPStransition.pdf, accessed June 6, 2017. And see Vaclav Smil, *Feeding the World: A Challenge for the Twenty-First Century* (Cambridge, MA: MIT Press, 2000).

24. Deborah Gewertz and Frederick Errington, *Cheap Meat: Flap Food Nations in the Pacific Islands* (Berkeley: University of California Press, 2010).

25. See Roger Horowitz, Jeffrey M. Pilcher, and Sydney Watts, "Meat for the Multitudes: Market Culture in Paris, New York City, and Mexico City over the Long Nineteenth Century," *American Historical Review* 109 (2004): 1055–1083.

26. For one example of this assumption being made by a prominent anthropologist of food, see Marvin Harris, *Good to Eat* (New York: Simon and Schuster, 1986).

27. See Johannes Fabian, *Time and the Other: How Anthropology Makes Its Object* (New York: Columbia University Press, 2014).

28. While some trace the paleo diet back to early twentieth-century antecedents, the first peer-reviewed article establishing some scientific basis for its benefits was S. Boyd Eaton and Melvin Konner, "Paleolithic Nutrition: A Consideration of Its Nature and Current Implications," *New England Journal of Medicine* 312 (1985): 283–289. The Paleolithic population that Eaton and Konner had in mind were those humans who lived in what is now Europe some forty thousand years ago. However, the contemporary nutritionist most often associated with the paleo diet is Loren Cordain. See Cordain et al., "Plant-Animal Subsistence Ratios and Macronutrient Energy Estimations in Worldwide Hunter-Gatherer Diets," *American Journal of Clinical Nutrition* 71 (2000): 682–692.

29. See Marion Nestle, "Paleolithic Diets: A Skeptical View," *Nutrition Bulletin* 25 (2000): 43–47.

30. For one version of this claim, see Marta Zaraska, *Meathooked: The History and Science of Our 2.5-Million-Year Obsession with Meat* (New York: Basic Books, 2016). Zaraska's source for this precise phrasing is the anthropologist Henry T. Bunn. See Bunn, "Meat Made Us Human," in Peter S. Ungar (ed.), *Evolution of the*

Human Diet (Oxford, UK: Oxford University Press, 2006). Notably, Zaraska's argument is not that meat was the *necessary* catalytic ingredient for producing the modern human condition, but that meat *happened to be* the particular "high-quality" food to which our ancestors had access to augment their diets.

31. For a skeptical treatment of what he terms "the hunting hypothesis," which addresses both human evolutionary origins and contemporary human behavior, see Matt Cartmill, *A View to a Death in the Morning* (Cambridge, MA: Harvard University Press, 1993).

32. See, for example, Craig B. Stanford's account of meat sharing among chimpanzees and gorillas: according to Stanford, not meat acquisition but meat sharing may have led to the development of a specific kind of social intelligence in our hominin ancestors. There is great value in Stanford's larger injunction to attend to mechanism; the evidence from both paleoanthropology and primate science implies that simple explanations are more likely to be misleading than useful. Stanford, *The Hunting Ape* (Princeton, NJ: Princeton University Press, 1999).

33. See Roger Lewin, *Human Evolution: An Illustrated Introduction* (Malden, MA: Blackwell, 2005).

34. See Leslie C. Aiello and Peter Wheeler, "The Expensive-Tissue Hypothesis: The Brain and the Digestive System in Human and Primate Evolution," *Current Anthropology* 36 (1995): 199–221.

35. But see Berson, *The Meat Question: Animals, Humans, and the Deep History of Food* (forthcoming), for a criticism of the argument that the brain could be fueled directly by the protein contained in meat.

36. See Ana Navarrete, Carel P. Van Schaik, and Karin Isler, "Energetics and the Evolution of Human Brain Size," *Nature* 480 (2011): 91–93.

37. Notably, Wrangham does credit the consumption of raw meat as having been an important driver of evolutionary change, pushing "our forebears out of the australopithecine rut," beginning, but not completing, the encephalization process and other physiological changes that produced *Homo sapiens*. See Wrangham, *Catching Fire: How Cooking Made Us Human* (New York: Basic Books, 2010), 103.

38. Donna Haraway, "The Past Is the Contested Zone," in *Simians, Cyborgs, and Women: The Reinvention of Nature* (New York: Routledge, 1991), 22. According to Haraway, the most important figure behind the "self-made species" argument is the primatologist Sherwood Washburn, regarded as one of the founders of his field; he was also one of the architects of the "man the hunter" idea.

39. See Gregory Schrempp, "Catching Wrangham: On the Mythology and the Science of Fire, Cooking, and Becoming Human," *Journal of Folklore Research* 48 (2011): 109–132.

40. E. O. Wilson, *Sociobiology: The New Synthesis* (Cambridge, MA: Harvard University Press, 1975). Wilson's book was not the first to advance sociobiological arguments; three years prior to its appearance, Lionel Tiger and Robin Fox's *Imperial Animal* (New York: Holt, Rinehart and Winston, 1972) was published. And sociobiology has scarcely quit the stage; in recent years, David Buss, Steven Pinker, and Yuval Harari have advanced what could be called sociobiological arguments,

from the precincts of evolutionary psychology and history, operating across very long timescales. As Ian Hesketh notes in his essay "The Story of Big History," the genre of "big" or extremely *longue durée* history often displays the trait of "consilience," or the desire for different bodies of evidence to lead to converging conclusions that are, in turn, productive of an overarching logic. As practiced by David Christian and others, "big history" seeks to place the stories of human civilization in a natural-historical context that greatly exceeds them in duration. See Hesketh, "The Story of Big History," *History of the Present* 4 (2014): 171–202. See also Martin Eger, "Hermeneutics and the New Epic of Science," in William Murdo McRae (ed.), *The Literature of Science: Perspectives on Popular Science Writing* (Athens: University of Georgia Press, 1993), 86–212. For a look at the role of "big history" in the more generalized use of historical materials in talks by entrepreneurs, see John Patrick Leary, "The Poverty of Entrepreneurship: The Silicon Valley Theory of History," *The New Inquiry,* June 9, 2017, https://thenewinquiry.com/the-poverty-of-entrepreneurship-the-silicon-valley-theory-of-history/, accessed June 11, 2017.

41. Mary Midgely, "Sociobiology," *Journal of Medical Ethics* 10 (1984): 158–160. She cites Wilson, *Sociobiology,* 4. See also Midgely, *Beast and Man: The Roots of Human Nature* (Brighton, UK: Harvester, 1978). For another view of sociobiology, see Howard L. Kaye, *The Social Meaning of Modern Biology: From Social Darwinism to Sociobiology* (New Haven, CT: Yale University Press, 1986).

42. See Peter Singer, "Ethics and Sociobiology," *Philosophy & Public Affairs* 11 (1982): 40–64; see, especially, p. 47. Singer quotes from Wilson, *Sociobiology,* 562.

43. Sahlins would, in terms of public profile, be eclipsed as a critic of sociobiology by biologists Richard Lewontin and Stephen Jay Gould. But for his original critique, see Marshall Sahlins, *The Use and Abuse of Biology: An Anthropological Critique of Sociobiology* (Ann Arbor: University of Michigan Press, 1976), 4.

44. The story of the always political culture-nature relation, in particular as instantiated in the debates over sociobiology, is too expansive both chronologically and synchronically for easy retelling here. For an early version, see Arthur Caplan's collection *The Sociobiology Debate* (New York: Harper & Row, 1978), which includes historical readings in the apparent nineteenth-century roots of sociobiological thinking, including Darwin and Spencer. See also W. R. Albury, "Politics and Rhetoric in the Sociobiology Debate," *Social Studies of Science* 10 (1980): 519–536. For a later telling, see Neil Jumonville, "The Cultural Politics of the Sociobiology Debate," *Journal of the History of Biology* 35 (2002): 569–593. For another examination of the field, one notably more charitable toward Wilson, see Ullica Segerstråle, *Defenders of the Truth: The Battle for Science in the Sociobiology Debate and Beyond* (Oxford, UK: Oxford University Press, 2001); for another, from the perspective of sociology, see Alexandra Maryanski, "The Pursuit of Human Nature by Sociobiology and by Evolutionary Sociology," *Sociological Perspectives* 37 (1994): 375–389. For another critical reading of sociobiology, which soundly rejects the idea of biology as a "nomothetic" foundation for the social sciences, see Lee Freese, "The Song of Sociobiology," *Sociological Perspectives* 37 (1994): 337–373.

45. On the independent complexity of the terms "culture" and "nature," see Raymond Williams, *Keywords: A Vocabulary of Culture and Society* (London:

Croom Helm, 1976). For one attack on the stability of the divide between culture and nature, made from the political Left and in the interests of the critique of industrial capitalism, see Haraway, "A Cyborg Manifesto," in Simians, Cyborgs, and Women.

46. Nor did Wilson invent the terminology of sociobiology, as he acknowledged; for a decades-earlier review of the state of the field, see G. Manoury, "Sociobiology," Synthese 5 (1947): 522–525. Intriguingly, one of the earliest "sociobiological" thinkers was Alexis Carrel, one of the founders of the discipline of tissue culture during the first decades of the twentieth century. He referred to his approach as "biosociology."

47. Notably, Howard L. Kaye argues against the view that sociobiology often seems to justify capitalism, claiming that sociobiology is less interested in naturalizing capitalism (and thus justifying it) and more interested in "reorder[ing] our psyches and societies." See Kaye, Social Meaning of Modern Biology, 5.

48. See Sahlins, Use and Abuse of Biology, 93.

49. Ibid., 100–102. For another version of this argument, see Haraway, "The Biological Enterprise: Sex, Mind, and Profit from Human Engineering to Sociobiology," in Simians, Cyborgs, and Women.

50. See Freese, "Song of Sociobiology," 345.

51. See Haraway, "The Past Is the Contested Zone: Human Nature and Theories of Production and Reproduction in Primate Behaviour Studies," in Simians, Cyborgs, and Women.

52. See Haraway, "Animal Sociology and a Natural Economy of the Body Politic: A Political Physiology of Dominance," in Simians, Cyborgs, and Women, 11.

53. See Marshall Sahlins, "The Original Affluent Society," in Stone Age Economics (Chicago, IL: Aldine-Atherton, 1972).

54. See William Laughlin, Richard B. Lee, and Irven deVore (eds.), with Jill Nash-Mitchell, Man the Hunter (Chicago, IL: Aldine-Atherton, 1968), 304. This book is based on the 1966 symposium of the same name.

55. See Lee and Devore, "Problems in the Study of Hunters and Gatherers," in Man the Hunter. However, note that Sherwood L. Washburn and C. S. Lancaster, in the same volume, seem to have strongly agreed. As they wrote, "In a very real sense, our intellect, emotions, and basic social life—all are evolutionary products of the success of the hunting adaptation." See Washburn and Lancaster, "The Evolution of Hunting," in Man the Hunter.

56. Perhaps the most important such claims, in terms of influence, were made by Justus von Liebig, the German chemist who, in his Animal Chemistry (1842), argued that protein was the only "true nutrient." On Liebig's influence, see William H. Brock, Justus von Liebig: The Chemical Gatekeeper (Cambridge, UK: Cambridge University Press, 1997). Liebig, notably, developed an idea with some similarities to cultured meat, in the mid-nineteenth century. In 1847 he published a method for making beef extract, with the intent of making a meat substitute that would address food insufficiencies around the world. Liebig would eventually participate in the establishment of a factory, in Uruguay, for the production of his beef extract, leading

to the creation of the London-based Liebig Extract of Meat Company in 1865; the company would later change its name to Oxo, which still makes beef bouillon cubes as of this writing.

57. See Michael S. Alvard and Lawrence Kuznar, "Deferred Harvests: The Transition from Hunting to Animal Husbandry," *American Anthropologist* 103 (2001): 295–311.

58. See Pat Shipman, "The Animal Connection and Human Evolution," *Current Anthropology* 51 (2010): 519–538.

59. Ibid., 524–525.

60. See also Helen M. Leach, "Human Domestication Reconsidered," *Current Anthropology* 44 (2003): 349–368.

61. See McGee, *On Food and Cooking,* 135.

62. See J. J. Harris, H. R. Cross, and J. W. Savell, "History of Meat Grading in the United States," Department of Animal Science, Texas A&M University, http://meat.tamu.edu/meat-grading-history/, accessed March 29, 2018.

63. On this subject, see, in particular, Orville Schell, *Modern Meat: Antibiotics, Hormones, and the Pharmaceutical Farm* (New York: Vintage, 1978).

64. William Boyd, "Making Meat: Science, Technology, and American Poultry Production," *Technology and Culture* 42 (2001): 631–664.

65. Friedrich Engels, *The Condition of the Working Class in England in 1844*, trans. Florence Kelley Wischnewetzky (London: George Allen & Unwin, 1892), 192.

66. See John Lossing Buck, "Agriculture and the Future of China," *Annals of the American Academy of Political and Social Science,* November 1, 1930.

67. On the importance of artificial fertilizers, see Vaclav Smil, "Population Growth and Nitrogen: An Exploration of a Critical Existential Link," *Population and Development Review* 17 (1991): 569–601.

68. The phrase is Siobhan Phillips's. See Phillips, "What We Talk about When We Talk about Food," *The Hudson Review* 62 (2009): 189–209; at 197.

69. See Zaraska, *Meathooked.*

70. William Cronon, *Nature's Metropolis: Chicago and the Great West* (New York: W. W. Norton, 1991), 256.

71. See Rachel Laudan, *Cuisine and Empire: Cooking in World History* (Berkeley: University of California Press, 2013), 208.

72. On the rise of sugar in British households, see Sidney Mintz, *Sweetness and Power: The Place of Sugar in Modern History* (New York: Viking, 1985).

73. Such reductions are usually associated with health-consciousness among wealthier eaters in the developed world. One report, produced by the Natural Resources Defense Council, shows that between 2005 and 2014, Americans reduced their beef consumption by about 20 percent; the report focuses on the desirable nature of this reduction, linked to the high carbon footprint of cattle-raising. See "Less Beef, Less Carbon," www.nrdc.org/sites/default/files/less-beef-less-carbon-ip.pdf, accessed June 6, 2017.

1. A third technique for duplicating animals, or their parts, in the laboratory should probably be on the table too: cloning from adult somatic cells that have been induced to return to a totipotent state, the technique used to create Dolly the sheep (1996–2003). On Dolly, as well as on the "potential" of cloning and transgenic techniques based on livestock cells, see Sarah Franklin, *Dolly Mixtures: The Remaking of Genealogy* (Durham, NC: Duke University Press, 2007).

2. See Shoshana Felman, *The Scandal of the Speaking Body: Don Juan with J. L. Austin, or Seduction in Two Languages* (Stanford, CA: Stanford University Press, 2003).

3. Mark Post, as quoted in "Lab-Grown Beef: 'Almost' Like a Burger," *Associated Press,* August 5, 2013.

4. This is from a statement by PETA, cited by Kate Kelland, "Scientists to Cook World's First In Vitro Beef Burger," *Reuters,* August 5, 2013, as well as by many other journalists covering Post's demonstration.

5. Jason Matheny, founder of New Harvest, quoted by Jason Gelt in "In Vitro Meat's Evolution," *The Los Angeles Times,* January 27, 2010.

6. I will meet this apple in late 2017, at a conference hosted by New Harvest in New York City. It is produced by Okanagan Specialty Fruits, which is owned by the Maryland-based biotech firm Intrexon. See Andrew Rosenblum, "GM Apples That Don't Brown to Reach U.S. Shelves This Fall," *MIT Technology Review,* October 7, 2017, www.technologyreview.com/s/609080/gm-apples-that-dont-brown-to-reach-us-shelves-this-fall/, accessed January 28, 2018.

7. Friedrich Nietzsche, *On the Genealogy of Morals,* trans. Walter Kaufmann (New York: Vintage Books, 1969), 57.

8. See Aristotle, *Politics* (Chicago, IL: University of Chicago Press, 2013).

9. Mike Fortun, *Promising Genomics: Iceland and deCODE Genetics in a World of Speculation* (Berkeley: University of California Press, 2008), 107.

10. See Hannah Arendt, *The Human Condition* (Chicago, IL: University of Chicago Press, 1958), 244–245.

11. Ibid., 245.

12. Ibid.

13. Arendt also compared sovereignty in the realm of politics to mastery in the realm of craftsmanship, or making, in an intriguing way: whereas sovereignty is effective because it is a communal and social practice, artistic mastery requires isolation. See ibid., 245.

14. See Merritt Roe Smith and Leo Marx (eds.), *Does Technology Drive History? The Dilemma of Technological Determinism* (Cambridge, MA: MIT Press, 1994).

15. See Merritt Roe Smith, "Technological Determinism in American Culture," in ibid. See also Roe Smith, "Technology, Industrialization, and the Idea of Progress in America," in K. B. Byrne (ed.), *Responsible Science: The Impact of Technology on Society* (Harper & Row, 1986), and Leo Marx, "Does Improved Technology Mean Progress?"

Technology Review (January 1987): 33–41, 71. The philosopher of science Langdon Winner has argued that, despite their Enlightenment origins, the belief in a link between technology and progress seems to have taken the place of a serious philosophy of technology or engineering. The idea of progress, Winner suggests, has kept us from achieving a deeper understanding of what we are doing with the devices that ease our way in this world. See Winner, *The Whale and the Reactor: The Search for Limits in the Age of High Technology* (Chicago, IL: University of Chicago Press, 1986), 5.

16. As Sarah Franklin points out, there is a link between the word "stem," as in stem cell, and the notion of "stock," as in genetic stock (i.e., the potential capital of livestock). See Franklin, *Dolly Mixtures,* 50, 57–58.

17. See Karen-Sue Taussig, Klaus Hoeyer, and Stefan Helmreich, "The Anthropology of Potentiality in Biomedicine," *Current Anthropology* 54, Supplement 7 (2013).

18. See Paul Martin, Nik Brown, and Alison Kraft, "From Bedside to Bench? Communities of Promise, Translational Research and the Making of Blood Stem Cells," *Science as Culture* 17 (2008): 29–41.

19. See Georges Canguilhem, "La théorie cellulaire," in Canguilhem, *La connaissance de la vie* (Paris: Vrin, 1989); translated into English by Stefanos Geroulanos and Daniela Ginsburg as "Cell Theory," in Canguilhem, *Knowledge of Life,* ed. Paola Marrati and Todd Meyers (New York: Fordham University Press, 2008), 43.

20. Isha Datar and Mirko Betti, "Possibilities for an in Vitro Meat Production System," *Innovative Food Science & Emerging Technologies* 11 (2010): 13–21.

21. I borrow this insight from Sarah Franklin. See Franklin, *Dolly Mixtures,* 59.

22. They fell under heavy criticism from other groups that hope to combat rhino poaching by more traditional means than biotechnology and market economics. See, for example, Katie Collins, "3D-Printed Rhino Horns Will Be 'Ready in Two Years'—but Could They Make Poaching Worse?" *Wired UK,* October 7, 2016, www.wired.co.uk/article/3d-printed-rhino-horns, accessed January 23, 2018.

23. For examples of anthropological works that explicitly deal with possible futures, see Lisa Messeri, *Placing Outer Space: An Earthly Ethnography of Other Worlds* (Chapel Hill, NC: Duke University Press, 2016); Ulf Hannerz, *Writing Future Worlds: An Anthropologist Explores Global Scenarios* (London: Palgrave, 2016); and Juan Francisco Salazar, Sarah Pink, Andrew Irving, and Johannes Sjöberg (eds.), *Anthropologies and Futures: Researching Emerging and Uncertain Worlds* (London: Bloomsbury, 2017). Another important anthropological angle on the future is the study of reproduction; see, in particular, Marilyn Strathern, *Reproducing the Future: Essays on Anthropology, Kinship and the New Reproductive Technologies* (Manchester, UK: Manchester University Press, 1992); and Strathern, "Future Kinship and the Study of Culture," *Futures* 27 (1995): 423–435. No account of anthropologists thinking about the future would be complete without mention of Margaret Mead's *The World Ahead: An Anthropologist Contemplates the Future* (New York: Berghahn, 2005).

24. See Johannes Fabian's *Time and the Other: How Anthropology Makes Its Object* (New York: Columbia University Press, 2014).

25. The notion of hot and cold societies, where "hot" connotes change and "cold" implies stability, comes from the don of structuralist anthropology, Claude Lévi-Strauss. For a reinterpretation of this notion, which also celebrates the cooler and plausibly more environmentally sustainable end of the temperature spectrum, see the essay by science fiction writer (and daughter of Alfred Louis Kroeber, the first professor of anthropology at the University of California, Berkeley) Ursula K. LeGuin, "A Non-Euclidean View of California as a Cool Place to Be," in *Dancing at the Edge of the World* (London: Gollancz, 1989).

CHAPTER FOUR. FOG

1. This is the one factual detail I have changed in order to serve this book's interpretive agenda: I actually saw the "Shareable Content" advertisement on a bus in Oakland in 2015, not in San Francisco in 2013.

2. Sigmund Freud, *Jokes and Their Relation to the Unconscious* (Standard Edition, vol. 8) (London: Hogarth Press, 1960), 118. On p. 146, Freud cites Herbert Spencer's "The Physiology of Laughter" (1860). Spencer had offered an "economic" model to describe the psychic energy that is discharged in laughter.

3. See Richard D. deShazo, Steven Bigler, and Leigh Baldwin Skipworth, "The Autopsy of Chicken Nuggets Reads 'Chicken Little,'" *American Journal of Medicine* 126 (2013): 1018–1019.

4. See Giovanni Arrighi, *The Long Twentieth Century: Money, Power, and the Origins of Our Times* (New York: Verso, 1994); see also the discussion of Arrighi in Raj Patel and Jason W. Moore, *A History of the World in Seven Cheap Things: A Guide to Capitalism, Nature, and the Future of the Planet* (Oakland: University of California Press, 2017), 69.

5. At the time of my visit to Breakout Labs, Thiel was a controversial public figure because of his outspoken libertarianism and his specific brand of futurism, described throughout this chapter. By the time of this book's writing, Thiel's name had become controversial for other, more explicitly political reasons, such as his vocal and financial support for conservative candidates for elected office, including the winner of the 2016 U.S. presidential election.

6. Quoted from the Breakout Labs website, www.breakoutlabs.org/, consulted July 4, 2017.

7. Modern Meadow would relocate to the Red Hook neighborhood of Brooklyn in 2014, with the intention of working more closely with the New York fashion industry. Modern Meadow would later withdraw from cultured meat production, focusing on biomaterials only. See chapter 10.

8. See George Packer, "No Death, No Taxes," *The New Yorker,* November 28, 2011.

9. For one example of journalistic coverage of the buses, see Casey Miner, "In a Divided San Francisco, Private Tech Buses Drive Tension," *All Tech Considered,*

December 17, 2013, www.npr.org/sections/alltechconsidered/2013/12/17/251960183/in-a-divided-san-francisco-private-tech-buses-drive-tension, accessed June 9, 2017.

10. See Nathan Heller, "California Screaming," *The New Yorker,* July 7, 2014.

11. Of course, there are far less dramatic ways to evict tenants, not all of them legal. The Anti-Eviction Mapping Project has made an effort to document illegal evictions in the Bay Area. See www.antievictionmappingproject.net/, accessed June 9, 2017.

12. A 2014 Brookings Institute study showed that income inequality had increased more dramatically in San Francisco than in any other U.S. city between 2007 and 2012. This was largely due to income gains for those in the 95th percentile of earners and above. See Alan Berube, "All Cities Are Not Created Unequal," February 4, 2014, www.brookings.edu/research/all-cities-are-not-created-unequal/, accessed June 9, 2017.

13. See Kenneth L. Kusmer, *Down and Out, On the Road: The Homeless in American History* (Oxford, UK: Oxford University Press, 2003). Kusmer writes of SoMa that it was once "the most important center of transient and casual labor on the West Coast."

14. See Meagan Day, "For More Than 100 Years, SoMa Has Been Home to the Homeless," https://timeline.com/for-more-than-100-years-soma-has-been-home-to-the-homeless-5e2d014bdd92, accessed June 17, 2017. The City of San Francisco conducts a "point in time count" to determine its homeless population; as of this writing, counts for recent years can be found at http://hsh.sfgov.org/research-reports/san-francisco-homeless-point-in-time-count-reports/.

15. See Day, "SoMa Has Been Home to the Homeless."

16. Cited in Stewart Brand, *The Clock of the Long Now: Time and Responsibility* (New York: Basic Books, 1999), 2–3.

17. On the interplay between science fictional artifacts and technology development in the actual world, see David A. Kirby, "The Future Is Now: Hollywood Science Consultants, Diegetic Prototypes and the Role of Cinematic Narratives in Generating Real-World Technological Development," *Social Studies of Science* 40 (2010): 41–70.

18. Years later, looking back at Breakout Labs' grantees, I see that they have been categorized under the following headings: "diagnostic," "therapy," "hardware," "cell bio," "nano" [nanotechnology], "synbio" [synthetic biology], "energy," "chemicals," "materials," "computing," "neuro," and "longevity." While none of these represents a specific problem the Thiel Foundation is trying to solve, it is easy to see that the last item, "longevity," could be interpreted to reflect a target condition rather than a straightforward category of technological application. See www.breakoutlabs.org/portfolio/, accessed August 2, 2017.

19. For a critical reading of the relationship between early twenty-first-century transhumanism and Silicon Valley capitalism, see Patrick McCray, "Bonfire of the Vainglorious," *Los Angeles Review of Books,* July 17, 2017, https://lareviewofbooks.org/article/silicon-valleys-bonfire-of-the-vainglorious/, accessed July 20, 2017.

CHAPTER FIVE. DOUBT

1. Pat Brown of Impossible Foods called lab-grown meat based on animal cells "one of the stupidest ideas ever expressed" in an interview published on May 22, 2017, at Techcrunch.com. See https://techcrunch.com/2017/05/22/impossible-foods-ceo-pat-brown-says-vcs-need-to-ask-harder-scientific-questions/, accessed November 7, 2017. I also draw here from remarks made by Brown at a talk given at Harvard University on September 21, 2017; I was in the audience.

2. Writer, chef, and television personality Anthony Bourdain said of "fake" meat that he sees it "as the enemy." See www.businessinsider.com/anthony-bourdain-big-problem-synthetic-fake-meat-laboratory-2016–12, accessed November 7, 2017. Bourdain addresses the problem of wasting parts of slaughtered animals in the video, suggesting that we should make much better use of the animal bodies we produce before we consider lab-grown meat.

3. Pew Research Center, "U.S. Views of Technology and the Future," April 17, 2014, http://assets.pewresearch.org/wp-content/uploads/sites/14/2014/04/US-Views-of-Technology-and-the-Future.pdf, accessed November 7, 2017.

4. Christina Agapakis, "Steak of the Art: The Fatal Flaws of In Vitro Meat," *Discover,* April 24, 2012.

5. See Warren Belasco, "Algae Burgers for a Hungry World? The Rise and Fall of Chlorella Cuisine," *Technology and Culture* 38 (1997): 608–634.

6. Christina Agapakis, "Growing the Future of Meat," *Scientific American,* August 6, 2013, https://blogs.scientificamerican.com/oscillator/growing-the-future-of-meat/, accessed October 21, 2017.

7. Ursula Franklin, *The Real World of Technology* (Toronto: Anansi Press, 1999).

CHAPTER SIX. HOPE

1. See Patrick D. Hopkins and Austin Dacey, "Vegetarian Meat: Could Technology Save Animals and Satisfy Meat Eaters?" *Journal of Agricultural and Environmental Ethics* 21 (2008): 579–596.

2. See Clemens Driessen and Michiel Korthals, "Pig Towers and In Vitro Meat: Disclosing Moral Worlds by Design," *Social Studies of Science* 42(2012): 797–820.

3. Neil Stephens and Martin Ruivenkamp, "Promise and Ontological Ambiguity in the *In Vitro* Meat Imagescape: From Laboratory Myotubes to the Cultured Burger," *Science as Culture* 25 (2016): 327–355.

4. See Nik Brown, "Hope against Hype—Accountability in Biopasts, Presents, and Futures," *Science Studies* 16(2) (2003): 3–21. See also Mike Fortun, *Promising Genomics: Iceland and deCODE Genetics in a World of Speculation* (Berkeley: University of California Press, 2008); and Kaushik Sunder Rajan, *Biocapital: The Constitution of Postgenomic Life* (Durham, NC: Duke University Press, 2006), 264.

5. See Fredric Jameson, *Archaeologies of the Future: The Desire Called Utopia and Other Science Fictions* (New York: Verso, 2005), 3.

CHAPTER SEVEN. TREE

1. The work in question was *Ice Arch* (1982). See Andy Goldsworthy Digital Catalogue, www.goldsworthy.cc.gla.ac.uk/image/?id=ag_02391, accessed July 14, 2017.

2. As it happened, Goldsworthy was working with the same Yorkshire stone used to pave the area around the de Young, a British import to San Francisco.

3. See Richard A. Walker, *The Country in the City: The Greening of the San Francisco Bay Area* (Seattle: University of Washington Press, 2007). See also Daegan Miller, *This Radical Land: A Natural History of American Dissent* (Chicago, IL: University of Chicago Press, 2018).

CHAPTER EIGHT. FUTURE

1. While bug entrepreneurs promote entomophagy, so do some policy experts. In 2013, the Food and Agriculture Organization (FAO) of the United Nations published a paper entitled "Edible Insects: Future Prospects for Food and Feed Security," in which they conclude that while hurdles do exist, "recent developments in research and development show edible insects to be a promising alternative for the conventional production of meat, either for direct human consumption or for indirect use as feedstock" (161). The FAO paper is available at www.fao.org/docrep/018/i3253e/i3253e.pdf. See also Julieta Ramos-Elorduy, "Anthropo-Entomophagy: Cultures, Evolution and Sustainability," *Annual Review of Entomology* 58 (2009): 141–160.

2. In 1971 the Consultative Group on International Agricultural Research appeared, followed by the Council for Agricultural Science and Technology. Food First, also known as the Institute for Food and Development Policy, was founded by Frances Moore Lappé in 1975. That same year, the International Food Policy Research Institute was founded in Washington, D.C. See Warren Belasco, *Meals to Come: A History of the Future of Food* (Berkeley: University of California Press, 2006), 55.

3. For an IFTF-produced timeline of major IFTF projects and concerns, see www.iftf.org/fileadmin/user_upload/images/whoweare/iftf_history_lg.gif, accessed February 5, 2019.

4. For an overview of futurists' methods, see Theodore J. Gordon, "The Methods of Futures Research," *Annals of the American Academy of Political and Social Science* 522 (1992): 25–35.

5. Fred Polak, "Crossing the Frontiers of the Unknown," in Alvin Toffler (ed.), *The Futurists* (New York: Random House, 1972).

6. See Simon Sadler, "The Dome and the Shack: The Dialectics of Hippie Enlightenment," in Iain Boal, Janferie Stone, Michael Watts, and Cal Winslow

(eds.), *West of Eden: Communes and Utopia in Northern California* (Oakland, CA: PM Press, 2012), 72–73. See also Fred Turner, *From Counterculture to Cyberculture: Stewart Brand, the Whole Earth Network, and the Rise of Digital Utopianism* (Chicago, IL: University of Chicago Press, 2006), 55–58.

7. "Introducing the IFTF Food Futures Lab," www.youtube.com/watch?v= 5_fP-7tfSK4, accessed November 1, 2017.

8. See, for example, the Center for Graphic Facilitation's homepage, http:// graphicfacilitation.blogs.com/pages/, accessed March 28, 2018.

9. The following account of mid- to late-twentieth-century futurism follows closely from Jenny Andersson, "The Great Future Debate and the Struggle for the World," *The American Historical Review* 117 (2012): 1411–1430; and Nils Gilman, *Mandarins of the Future: Modernization Theory in Postwar America* (Baltimore, MD: Johns Hopkins University Press, 2003).

10. See Olaf Helmer, "Science," *Science Journal* 3(10) (1967): 49–51, 51.

11. See T.J. Gordon and Olaf Helmer, "Report on a Long-Range Forecasting Study" (Santa Monica, CA: Rand, 1964).

12. This ideology-vs.-rationalist typology is Jenny Andersson's.

13. Walt Whitman Rostow, *The Stages of Economic Growth: A Non-Communist Manifesto* (Cambridge, UK: Cambridge University Press, 1960), 2.

14. See Daniel Bell and Steven Graubard, *Toward the Year 2000: Work in Progress,* special issue of *Daedalus* (1967), republished by MIT Press in 1969.

15. Daniel Bell, *The Coming of Post-industrial Society: A Venture in Social Forecasting* (New York: Basic Books, 1973).

16. See Gilman, *Mandarins of the Future.*

17. Brendan Buhler, "On Eating Roadkill, The Most Ethical Meat," *Modern Farmer,* September 12, 2013, http://modernfarmer.com/2013/09/eating-roadkill /, accessed August 24, 2017. The first in-depth examination of roadkill was James R. Simmons, *Feathers and Fur on the Turnpike* (Boston: Christopher, 1938).

18. Allan Savory's claims on behalf of holistic grasslands management, which include that it can help combat global warming by bringing back carbon-sequestering grasslands, have not gone uncriticized. See James E. McWilliams, "All Sizzle and No Steak," *Slate,* April 22, 2013, www.slate.com/articles/life/food/2013/04/allan _savory_s_ted_talk_is_wrong_and_the_benefits_of_holistic_grazing_have .html, accessed August 24, 2017.

19. My account of the place of meat in food futurism follows closely from Warren Belasco, *Meals to Come.*

20. See Thomas Robert Malthus, *An Essay on the Principle of Population* (London: Penguin, 1985), 187–188.

21. On the influence of improvement efforts in British agriculture on Malthus, see Fredrik Albritton Jonsson, "Island, Nation, Planet: Malthus in the Enlightenment," in Robert Mayhew (ed.), *New Perspectives on Malthus* (Oxford, UK: Oxford University Press, forthcoming).

22. Paul and Anne Ehrlich, *The Population Bomb* (New York: Ballantine, 1968). See also Thomas Robertson, *The Malthusian Moment: Global Population Growth*

and the Birth of American Environmentalism (New Brunswick, NJ: Rutgers University Press, 2012).

23. See Paul and Anne Ehrlich, *One with Nineveh: Politics, Consumption, and the Human Future* (Washington, DC: Island Press, 2004). And see Paul Ehrlich's interview in the *Los Angeles Times,* in which he discusses what he feels *The Population Bomb* got right and wrong: Patt Morrison, "Paul R. Ehrlich: Saving Earth. The Scholar Looks the Planet, and Humanity, in the Face," February 12, 2011, http://articles.latimes .com/2011/feb/12/opinion/la-oe-morrison-ehrlich-021211, accessed September 13, 2017.

24. Fredrik Albritton Jonsson, "The Origins of Cornucopianism: A Preliminary Genealogy," *Critical Historical Studies* 1 (2014): 151–168.

25. Albritton Jonsson also provides a wide array of precursors to Ricardo's political-economic crystallization of cornucopianism. To list just a few: Francis Bacon's natural philosophy; the popular Newtonianism of the eighteenth century, as expressed through civil engineering projects and the effort of Charles Webster and Thomas Hughes to "recover Eden" through technological progress; and the experience of North American colonization itself, in which expansion and New World economic activity seemed to implicitly promise future abundance. Ibid.

26. In the fall of 2015, I found myself at a conference ostensibly focused on how to balance ecological concerns with business interests. One of the keynote speakers, however, was instead focused on the question of how using satellites to mine asteroids for natural resources might help us "support the permanent expansion of human settlement and economic activity beyond our earth," as he put it. Ricardo's views have many distant descendants.

27. For example, see Joel Mokyr, *The Gift of Athena: Historical Origins of the Knowledge Economy* (Princeton, NJ: Princeton University Press, 2002).

28. Albritton Jonsson, "Origins of Cornucopianism," 160.

29. See, for example, Will Steffen, Paul Crutzen, and John McNeill, "The Anthropocene: Are Humans Now Overwhelming the Great Forces of Nature?" *AMBIO: A Journal of the Human Environment* 36 (2007): 849–852.

30. See Gilman, *Mandarins of the Future,* 1–2.

31. Karl Marx, *The Communist Manifesto: With Related Documents* (Boston: Bedford/St. Martin's, 1999).

32. See Marshall Berman, *All That Is Solid Melts into Air: The Experience of Modernity* (New York: Verso, 1982), 21. My analysis of this passage from Marx derives from Berman's.

CHAPTER NINE. PROMETHEUS

1. See Gregory Schrempp, "Catching Wrangham: On the Mythology and the Science of Fire, Cooking, and Becoming Human," *Journal of Folklore Research* 48 (2011): 109–132.

2. There are many important studies of the myth of Prometheus. See, for example, Hans Blumenberg, *Arbeit am Mythos* (Frankfurt am Main: Suhrkamp, 1979),

in English translation by Robert M. Wallace as *Work on Myth* (Cambridge, MA: MIT Press, 1985); and Raymond Trousson, *Le thème de Prométhée dans la littérature européenne* (Geneva: Librairie Droz, 1964). See also Alfredo Ferrarin, "Homo Faber, Homo Sapiens, or Homo Politicus? Protagoras and the Myth of Prometheus," *The Review of Metaphysics* 54 (2000): 289–319. Ferrarin's study is notable for its use of the Prometheus story in order to reach a new appreciation of Greek understandings of *techne,* or skill in making.

3. Schwartz's original reads, "Biology itself is invested with the rich ambivalence of myth." See Hillel Schwartz, *The Culture of the Copy: Striking Likenesses, Unreasonable Facsimiles* (revised and updated) (New York: Zone Books, 2014), 19.

4. Gaston Bachelard, *The Psychoanalysis of Fire,* trans. Alan C.M. Ross (London: Routledge & Kegan Paul, 1964).

5. On the premodern reception history of Prometheus's story, see Olga Raggio, "The Myth of Prometheus: Its Survival and Metaphormoses up to the Eighteenth Century," *Journal of the Warburg and Courtauld Institutes* 21 (1958): 44–62.

6. Mary Shelley's novel of 1819, *Frankenstein,* bears the subtitle "The Modern Prometheus," a reference to her protagonist Victor Frankenstein, often understood to be a stand-in for her husband Percy. Sometimes called the first science fiction novel, *Frankenstein* is also a story of flesh animated through artifice, written before the advent of modern cell theory. In the very early nineteenth century, cellular biology had not yet killed off the idea that life might be a principle of power rather than an effect produced by the interplay of biological parts. As the literary scholar Denise Gigante argues, *Frankenstein* may be read as exploring this idea of life as power, one of a number of products of literary Romanticism that do so. *Frankenstein* does not answer the question of life itself but explores its protagonist's obsession with that question, implying that by creating an artificial "double," Victor Frankenstein was loosing upon the world something unpredictable, a form of life that would inevitably exceed not only the limits of mere matter, but also the morality of its maker. But Frankenstein's monster seems to have had his own morality. He was, perhaps in reference to Percy, a vegetarian, given that he says, "My food is not that of man; I do not destroy the lamb and the kid, to glut my appetite; acorns and berries afford me sufficient nourishment." Mary Shelley, *Frankenstein: The Modern Prometheus* (London: Henry Colburn and Richard Bentley, 1831), 308. And see Denise Gigante, *Life: Organic Form and Romanticism* (New Haven, CT: Yale University Press, 2009), especially pp. 1–48, 160–163. See also Carol Adams's discussion of Frankenstein as a vegetarian in *The Sexual Politics of Meat: A Feminist-Vegetarian Critical Theory* (New York: Continuum, 1990), 108–119.

CHAPTER TEN. MEMENTO

1. Arthur Schopenhauer, *The World as Will and Representation,* trans. and ed. Judith Norman, Alistair Welchman, and Christopher Janaway (Cambridge, UK: Cambridge University Press, 2010).

2. See Heather Paxson, *The Life of Cheese: Crafting Food and Value in America* (Berkeley: University of California Press, 2012), 31.

3. See Rachel Laudan, "A Plea for Culinary Modernism: Why We Should Love New, Fast, Processed Food," *Gastronomica* 1 (2001): 36–44.

4. Ibid., 36.

5. Tim Lang, a professor of food policy at City University in London, coined the term "food miles." The concept was featured in his publication "The Food Miles Report: The Dangers of Long-Distance Food Transportation," SAFE Alliance, 1994. Criticisms of the "food miles" approach to assessing the environmental impact of food systems usually focus on the fact that transportation often comprises only a small fraction of the impact of the food system, with a greater portion of a food's "carbon footprint" resulting from production. See, for example, Pierre Desrochers and Hiroko Shimizu, "Yes, We Have No Bananas: A Critique of the 'Food-Miles' Perspective," George Mason University Mercatus Policy Series, Policy Primer No. 8, October 2008.

6. See William Cronon, *Nature's Metropolis: Chicago and the Great West* (New York: W. W. Norton, 1991).

7. As Clemens Driessen and Michiel Korthals have shown, cultured meat imaginaries derive inspiration from urban agriculture. See their "Pig Towers and In Vitro Meat: Disclosing Moral Worlds by Design," *Social Studies of Science* 42 (2012): 797–820.

8. For a brief history of the Gowanus Canal's designation as a Superfund site, see Juan-Andres Leon, "The Gowanus Canal: The Fight for Brooklyn's Coolest Superfund Site," in *Distillations,* winter 2015, www.chemheritage.org/distillations/magazine/the-gowanus-canal, accessed March 6, 2017.

9. Proteus Gowanus was still open during the events this chapter describes but concluded its ten years of operation just afterwards, in 2015.

10. For an anthropological study of raising meat animals that takes "authenticity," "locality," and "terroir" as central terms for interrogation, see Brad Weiss, *Real Pigs: Shifting Values in the Field of Local Pork* (Durham, NC: Duke University Press, 2016).

11. See Max Weber, *The Protestant Ethic and the Spirit of Capitalism,* trans. Talcott Parsons (New York: Routledge, 2001). For one variegated reading of artisanal food production and value, and of the ways commercial activities around food are understood to produce extra-commercial forms of value, see Paxson, *Life of Cheese.*

12. The idea that cows release a problematically high volume of methane through their burps is not without its critics. According to Nicolette Hahn Niman (notably, a former attorney married to Bill Niman, the founder of Niman Ranch, a San Francisco Bay Area meat producer specializing in transparency, humanely raised animals, and environmental sustainability), the 14–18 percent figure often heard is based on studies of methane production in cattle that employ very small sample sizes. Therefore, it is inappropriate to extrapolate from this to describe the global population of cattle. See Nicolette Hahn Niman, *Defending Beef: The Case For Sustainable Meat Production* (White River Junction, VT: Chelsea Green, 2014).

13. Nathan Heller, "Listen and Learn," *The New Yorker,* July 9, 2012.

14. Catherine Mohr, "Surgery's Past, Present, and Robotic Future," TED 2009, www.ted.com/talks/catherine_mohr_surgery_s_past_present_and_robotic_future#t-1100302, accessed November 14, 2017.

15. Wurman founded TED in 1984 and sold it in 2003, going on to found other, related conferences, in some of which he rethought the format. At the same time, the new director of TED, Chris Anderson, transformed the conference into the cultural phenomenon it would become in the years that followed.

16. See Laudan, "Plea for Culinary Modernism," 43.

CHAPTER ELEVEN. COPY

1. See Walter Benjamin, "The Work of Art in the Age of Mechanical Reproduction," in *Illuminations: Essays and Reflections,* trans. Harry Zohn (New York: Harcourt Brace Jovanovich, 1968).

2. On the cultural history of the repetition of experiences, from déjà vu to the philosophical views of Kierkegaard or Nietzsche, see Hillel Schwartz, *The Culture of the Copy: Striking Likenesses, Unreasonable Facsimiles* (revised and updated) (New York: Zone Books, 2014).

3. Benjamin, *Illuminations,* 188.

4. See Corby Kummer, *The Pleasures of Slow Food* (San Francisco: Chronicle Books, 2002).

5. See Laudan, "A Plea for Culinary Modernism: Why We Should Love New, Fast, Processed Food," *Gastronomica* 1 (2001): 36–44.

6. The idea of the importance of biological equivalency is where cultured meat splits off from an older tradition of copying meat, one based on what we might call sensory equivalency. For a discussion of this point, and of efforts to produce sensory-equivalent meat surrogates based on plant protein, see my "Meat Mimesis: Laboratory-Grown Meat as a Study in Copying," forthcoming in *Osiris.*

7. See Schwartz, *Culture of the Copy,* for one of the most important surveys of the subject.

8. Hans Blumenberg, "Imitation of Nature: Toward a Prehistory of the Idea of the Creative Being [first published in 1957]," trans. Ania Wertz, *Qui Parle* 12(1) (2000): 17–54.

9. As Blumenberg points out, the Aristotelian position on mimesis was already a response to Plato, and in particular to a question within Platonism having to do with the origin of human works. Were there Forms for artificial objects, as Plato seems to imply in Book 10 of the *Republic?* Plato's Academy appears to have dropped such a notion by Aristotle's time, replacing it with the idea that the cosmos built upon the Forms reflects the best of that which is to come, and that there are no "leftover" Forms human artisans might tap (see the *Timaeus*). In summary, the Aristotelian response is to deny the existence of invention that is not imitation of nature. See Blumenberg, "Imitation of Nature," 29.

10. For certain followers of Aristotle, the distinction between *natura naturans* and *natura naturata* took on a sexed dimension, as the former was understood to be a male productive principle, while the finished but inert produce was taken to be female. This was an adaptation of Aristotle's own view that the mother's body simply provides the raw material for the process of reproduction. See Mary Garrard, "Leonardo da Vinci: Female Portraits, Female Nature," in Mary Garrard and Norma Broude (eds.), *The Expanding Discourse: Feminism and Art History* (New York: IconEditions, 1992), 58–86.

CHAPTER TWELVE. PHILOSOPHERS

1. See Patrick Martins with Mike Edison, *The Carnivore Manifesto: Eating Well, Eating Responsibly, and Eating Meat* (New York: Little, Brown, 2014).

2. For this writing I consulted the revised 2002 edition: Peter Singer, *Animal Liberation* (New York: HarperCollins, 2002). Notably, the book's reputation as a bible of the animal rights movement gets its philosophical argument, which is utilitarian rather than rights based, badly wrong. In the book, Singer states that its argument cannot be refuted by counterarguments to the effect that animals do not possess rights and that "the language of rights" is merely "a convenient political shorthand." Ibid., 8. Elsewhere, Singer would regret his "concession to popular moral rhetoric" on the grounds that it allowed critics to confuse his argument for a rights theory; see Singer, "The Fable of the Fox and the Unliberated Animals," *Ethics* 88 (1978): 119–125; see 122.

3. Singer, "Utilitarianism and Vegetarianism," in *Philosophy & Public Affairs* 9 (1980): 325–337.

4. Collectively Free have posted their own narrative account of their action at the panel, with video, online: www.collectivelyfree.org/intervention-4-all-animals-want-to-live-museum-of-food-and-drink-mofad-manhattan-ny/, accessed March 13, 2018.

5. For an example of Budolfson's work, including his criticisms of Singer's version of utilitarianism, see Mark B. Budolfson, "Is It Wrong to Eat Meat from Factory Farms? If So, Why?" in Ben Bramble and Bob Fischer (eds.), *The Moral Complexities of Eating Meat* (Oxford, UK: Oxford University Press, 2015). Some of Budolfson's work, including this essay, is devoted to the problem of structural complicity in harm and acknowledges that the vast majority of consumer behavior in the developed world is connected, in one way or another, to environmentally or ethically problematic behavior.

6. Bernard Williams, "A Critique of Utilitarianism" in J. J. C. Smart and Bernard Williams, *Utilitarianism: For and Against* (Cambridge, UK: Cambridge University Press, 1973), 137.

7. Bertrand Russell, "The Harm That Good Men Do," *Harpers,* October, 1926. It is important not to take Russell's crediting of utilitarianism at face value; he had a personal interest in a positive review of its historical contribution.

8. See Michel Foucault, "Truth and Juridical Forms," in *Power: Essential Works of Foucault, 1954–1984*, ed. Paul Rabinow (New York: The New Press, 2000), 70.

Note that I borrow the device of pairing Russell's and Foucault's varying accounts of utilitarianism from Bart Schultz and Georgios Varouxakis (eds.), *Utilitarianism and Empire* (Lanham, MD: Lexington Books, 2005); see the introduction.

9. See Jeremy Bentham, "A Comment on the Commentaries and A Fragment on Government," ed. J. H. Burns and H. L. A. Hart, in *The Collected Works of Jeremy Bentham* (Oxford, UK: Oxford University Press, 1970), 393.

10. Alasdair MacIntyre in conversation with Alex Voorhoeve, *Conversations on Ethics* (Oxford, UK: Oxford University Press, 2009), 116.

11. For a brief account of Bentham's desires for influence see James Crimmins, "Bentham and Utilitarianism in the Early Nineteenth Century," in B. Eggleston and D. Miller (eds.), *The Cambridge Companion to Utilitarianism* (New York: Cambridge University Press, 2014), 38. For a more extended account, see Crimmins, *Secular Utilitarianism: Social Science and the Critique of Religion in the Thought of Jeremy Bentham* (Oxford, UK: Oxford University Press, 1990).

12. See Christine M. Korsgaard, "Getting Animals in View," *The Point,* no. 6, 2013.

13. Ibid., 123.

14. Paul Muldoon, "Myself and Pangur," in *Hay* (New York: Farrar, Straus and Giroux, 1998). The poem is an adaptation of an oft-translated, anonymous poem thought to have been written by an Irish monk in the ninth century C.E.

15. This is a point on which Korsgaard, who is in many ways a Kantian, converges with Singer, a surprising turn given that Kantian—or, more broadly, deontological—reasoning is often juxtaposed against consequentialism as if they were opposites. Korsgaard writes, "The claim of the other animals to the standing of ends in themselves has the same ultimate foundation as our own—the essentially self-affirming nature of life itself." Korsgaard, "Getting Animals in View."

16. See Singer, *The Expanding Circle: Ethics, Evolution, and Moral Progress* (Princeton, NJ: Princeton University Press, 2011). In this book Singer sets an old idea, the moral circle and its expansion via altruistic behaviors, in sociobiological terms. The notion of a moral circle, also called a "circle of concern" or "circle of moral regard," which relates to our capacities for sympathy, empathy, and altruism, has a recorded history in Western philosophy that goes back to Aristotle. Aristotle writes in the *Eudemian Ethics:*

> As to seeking for ourselves and praying for many friends, and at the same time saying that one who has many friends has no friend, both statements are correct. For if it is possible to live with and share the perceptions of many at once, it is most desirable for them to be the largest possible number; but as that is very difficult, active community of perception must of necessity be in a smaller circle, so that it is not only difficult to acquire many friends (for probation is needed), but also to use them when one has got them. (1245b17–18)

Hierocles the Stoic, in his second-century C.E. *Elements of Ethics,* describes a set of concentric rings of relation beginning with the self and extending out through levels of increasingly distant familial relatedness. See Ilaria Ramelli, *Hierocles the*

Stoic: Elements of Ethics, Fragments, and Excerpts (Leiden, The Netherlands: Brill, 2009), 91–93. As Martha Nussbaum argues, Greek thought allowed for the expansion of moral concern and explains it through the faculty of the imagination; for the ancient Greeks, it was often clear that while there are differences of rank between persons, "many of the most important distinctions among human beings are the work of fortune, unconnected to human desert." See Nussbaum, "Golden Rule Arguments: A Missing Thought?" in Kim-chong Chong, Sor-hoon Tan, and C. L. Ten (eds.), *The Moral Circle and the Self: Chinese and Western Approaches* (Chicago, IL: Open Court, 2003), 9. Nussbaum points out that there are many modern versions of the Greek idea in Rousseau and others, and we find it in such contemporary philosophers as John Rawls. However, the notion of a historical march of moral progress expressed as the expansion of a circle appears to be modern in origin. See, for example, William Edward Hartpole Lecky, *History of European Morals from August to Charlamagne,* vol. 1 (London: Longmans, Green, 1890), 107. The salient issue for Singer, however, is not the expansion of the moral circle among humans but the imaginative enlargement of that circle to include other creatures as well. It was Charles Darwin who claimed to explain interspecies altruism on a natural basis, and to identify such natural sympathy ("to suffer with," from the Greek) as a key to the development of civilization. On sympathy in Darwin and in Victorian culture, see Rob Boddice, *The Science of Sympathy: Morality, Evolution, and Victorian Civilization* (Urbana: University of Illinois Press, 2016).

17. Immanuel Kant, "Conjectures on the Beginning of Human History," cited in Korsgaard, "Getting Animals in View" (the ellipses are Korsgaard's; the capitalization of "Man" is mine); see the full text in H. S. Reiss (ed.), *Kant: Political Writings* (Cambridge, UK: Cambridge University Press, 1970).

18. See Jeremy Bentham, "Introduction to the Principles of Morals and Legislation," in *Collected Works of Jeremy Bentham.*

19. And see Ryder's essay "Experiments on Animals" in *Animals, Men and Morals: An Enquiry into the Maltreatment of Non-humans* (London: Gollancz, 1971); for "speciesism," established by analogy with racism, see p. 81. Ryder quotes "the late Professor C. S. Lewis" as follows: "If loyalty to our own species, preference for man simply because we are men, is not a sentiment, then what is? If mere sentiment justifies cruelty, why stop at a sentiment for the whole human race? There is also a sentiment for a white man against the black, for Herrenvolk against the Non-Aryans."

20. But note that Tom Regan, one of Singer's critics (about whom more below), has noted that Singer failed to provide a strictly utilitarian argument against speciesism, instead presenting what Regan terms an argument for moral consistency for its own sake. I note that an argument for moral consistency for its own sake comes rather close to a deontological position, which is the opposite of what one would expect from Singer. Regan takes his argument further, saying that Singer has in fact failed to show that there is a utilitarian argument on behalf of vegetarianism, something that in his view would require a mass of empirical data that Singer fails to

provide. See Regan, "Utilitarianism, Vegetarianism, and Animal Rights," *Philosophy & Public Affairs* 9 (1980): 305–324.

21. For some early critiques of Singer, see, for example, Michael Martin, "A Moral Critique of Vegetarianism," *Reason Papers,* no. 3 (1976), 13–43; Philip Devine, "The Moral Basis of Vegetarianism, *Philosophy* 53 (1978), 481–505; Leslie Pickering Francis and Richard Norman, "Some Animals Are More Equal Than Others," *Philosophy* 53 (1978), 507–527; Aubrey Townsend, "Radical Vegetarians," *Australasian Journal of Philosophy* 57 (1979), 85–93; Peter Wenz, "Act-Utilitarianism and Animal Liberation," *The Personalist* 60 (1979): 423–428; and R.G. Frey, *Rights, Killing, and Suffering: Moral Vegetarianism and Applied Ethics* (Oxford, UK: Blackwell, 1983). For a more recent assortment of attacks on Singer, with a response from Singer, see Jeffrey A. Schaler (ed.), *Peter Singer Under Fire: The Moral Iconoclast Faces His Critics* (Chicago, IL: Open Court, 2009).

22. See Gary L. Francione, "On Killing Animals," in *The Point,* no. 6 (2013); and Francione, *Animals as Persons: Essays on the Abolition of Animal Exploitation* (New York: Columbia University Press, 2008).

23. See Tom Regan, "The Moral Basis of Vegetarianism," *Canadian Journal of Philosophy* 5 (1975): 181–214; Regan, "Utilitarianism, Vegetarianism, and Animal Rights," *Philosophy & Public Affairs* 9 (1980), 305–324; and Regan, *The Case for Animal Rights* (Berkeley: University of California Press, 1983).

24. See Michael Fox, "'Animal Liberation': A Critique," *Ethics* 88 (1978): 106–118.

25. See responses to Fox by Singer, "Fable of the Fox," and by Regan, "Fox's Critique of Animal Liberation," *Ethics* 88 (1978): 126–133. Fox responded to both Singer and Regan in the same issue of *Ethics:* see "Animal Suffering and Rights: A Reply to Singer and Regan," *Ethics* 88 (1978): 134–138.

26. "Every moving thing that lives shall be food for you; and just as I gave you the green plants, I give you everything." *New Oxford Annotated Bible,* New Standard Revised Version, ed. Bruce M. Metzger and Roland E. Murphy (New York: Oxford University Press, 1991), 12.

27. Readers familiar with David Foster Wallace's essay "Consider the Lobster," which reaches the same conclusion about the weaknesses of moral-philosophical defenses of eating animals as this chapter, have at this point collected all the Easter eggs I have hidden in reference to that essay and can relax. See Wallace, "Consider the Lobster," *Gourmet,* August 2004, 50–64.

CHAPTER THIRTEEN. MAASTRICHT

1. For a general guide to the landscape of the Netherlands, see Audrey M. Lambert, *The Making of the Dutch Landscape: An Historical Geography of the Netherlands* (London: Academic Press, 1985). On the history of dikes and polders in the Netherlands, see Eric-Jan Pleijster, *Dutch Dikes* (Rotterdam: nai010, 2014).

2. The Dutch are, on average, the world's tallest people. And the history of Dutch height contains a mystery: why have Dutch men gained some twenty centimeters (7.87 inches) in height over the past two hundred years? Military records indicate that in the mid-eighteenth century, Dutch men were short compared to other Europeans, and that they subsequently shot up beyond their peers. One team of researchers working with both historical records and contemporary biometric data has concluded that the Dutch height gain was congruent with stature increases across Western populations, gains plausibly caused by improved diets during a period of democratized access to nutritionally rich foods such as milk, eggs, fish, and meat. The Dutch gain just kept going after other nationalities leveled off. Possible reasons include continued improvements to the Dutch diet, as well as natural selection favoring mothers of average height and tall fathers. See Gert Stulp et al., "Does Natural Selection Favour Taller Stature among the Tallest People on Earth?" *Proceedings of the Royal Society B* 282 (2015): 20150211.

3. See Rachel Laudan, "A Plea for Culinary Modernism: Why We Should Love New, Fast, Processed Food," *Gastronomica* 1 (2001): 36–44.

4. On Harrison's *Make Room! Make Room!* and Malthusianism, see Warren Belasco, *Meals to Come: A History of the Future of Food* (Berkeley: University of California Press, 2006), 51, 134. Paul Ehrlich, author of *The Population Bomb,* wrote an introduction for the paperback of *Make Room! Make Room!*

5. Hans Blumenberg, *Care Crosses the River,* trans. Paul Fleming (Stanford, CA: Stanford University Press, 2010), 133.

6. For notes on Maastricht's geological and political history from which I have borrowed liberally, see John McPhee, "A Season on the Chalk," in *Silk Parachute* (New York: Macmillan, 2010).

7. I am, in fact, such an urban person that I can count the times I have met the gaze of an animal who was not a house pet using only my fingers and toes. John Berger writes that "a power is ascribed to the animal, comparable with human power but never coinciding with it. The animal has secrets which, unlike the secrets of caves, mountains, seas, are specifically addressed to man." Berger, "Why Look at Animals?" in *About Looking* (New York: Vintage, 1991).

8. On Roman Maastricht, see Lambert, *Making of the Dutch Landscape.*

9. See Sheila Jasanoff, *Designs on Nature: Science and Democracy in Europe and the United States* (Princeton, NJ: Princeton University Press, 2007).

10. See Cor van der Weele and Clemens Driessen, "Animal Liberation?" https://bistro-invitro.com/en/essay-cor-van-der-weele-animal-liberation/, accessed January 11, 2018; and van der Weele and Driessen, "Emerging Profiles for Cultured Meat; Ethics through and as Design," *Animals* 3 (2013): 647–662. See also Wim Verbeke, Pierre Sans, and Ellen J. Van Loo, "Challenges and Prospects for Consumer Acceptance of Cultured Meat," *Journal of Integrative Agriculture* 14, (2015): 285–294; and Verbeke et al., "Would You Eat 'Cultured Meat'?: Consumers' Reactions and Attitude Formation in Belgium, Portugal and the United Kingdom," *Meat Science* 102 (2015): 49–58.

11. See McPhee, "Season on the Chalk."

1. She was referring to a discussion thread that New Harvest had held at Reddit .com in March 2016: www.reddit.com/r/IAmA/comments/48sn01/we_are_new _harvest_the_nonprofit_responsible_for/, accessed March 28, 2018.

2. Mary Douglas, *Purity and Danger* (London: Routledge, 1966), 51.

3. Some animal protection activists opposed to kosher slaughter argue that it is no less cruel than nonkosher slaughter. Additionally, many activists assert that kosher slaughter, when conducted at an industrial scale, leads to conditions that produce considerable suffering for animals. Bruce Friedrich (as of this writing, head of the Good Food Institute; previously at PETA) once debated attorney Nathan Lewin on this point; a recording of part of that debate is available at www.mediapeta.com /peta/Audio/bruce_debate_final.mp3, accessed October 14, 2018. Temple Grandin, a professor of animal science and noted expert on slaughter practices, has written a balanced editorial about the ethical problems facing kosher slaughter at an industrial scale, in an American Jewish newspaper: Grandin, "Maximizing Animal Welfare in Kosher Slaughter," *The Forward,* April 27, 2011, http://forward.com/opinion/137318 /maximizing-animal-welfare-in-kosher-slaughter/, accessed October 14, 2018.

4. See Shmuly Yanklowitz, "Why This Rabbi Is Swearing Off Kosher Meat," *The Wall Street Journal,* May 30, 2014.

5. Sarah Zhang, "A Startup Wants to Grow Kosher Meat in a Lab," *The Atlantic,* September 16, 2016, www.theatlantic.com/health/archive/2016/09/is-lab-grown-meat-kosher/500300/, accessed October 14, 2018.

6. See Roger Horowitz, *Kosher USA: How Coke Became Kosher and Other Tales of Modern Food* (New York: Columbia University Press, 2016).

7. Ibid., 125.

CHAPTER FIFTEEN. WHALE

1. For a sharply critical analysis of the use of historical materials by speakers and authors addressing business audiences, see John Patrick Leary, "The Poverty of Entrepreneurship: The Silicon Valley Theory of History," *The New Inquiry,* June 9, 2017.

2. After the event ended, Shapiro mentioned the source for his version of the story: Eric Jay Dolin's *Leviathan: The History of Whaling in America* (New York: W. W. Norton, 2007).

3. For a similar but thinner telling, see Amory B. Lovins, "A Farewell to Fossil Fuels: Answering the Energy Challenge," *Foreign Affairs* 91 (2012): 134–146.

4. See Bill Kovarik, "Thar She Blows! The Whale Oil Myth Surfaces Again," TheDailyClimate.org, March 3, 2014; and Kovarik, "Henry Ford, Charles Kettering, and the Fuel of the Future," *Automotive History Review* (spring 1998): 7–27.

5. See Dan Bouk and D. Graham Burnett, "Knowledge of Leviathan: Charles W. Morgan Anatomizes His Whale," *Journal of the Early Republic* 28 (2008): 433–466.

6. On whale oil as a resource extracted faster than it is replenished, in such a way as to display what economists call a Hubbert curve, not unlike petroleum, see U. Bardi, "Energy Prices and Resource Depletion: Lessons from the Case of Whaling in the Nineteenth Century," *Energy Sources B* 2 (2007): 297–304.

CHAPTER SIXTEEN. CANNIBALS

1. See Mary Douglas, *Purity and Danger* (London: Routledge, 1966).

2. Sigmund Freud, "From the History of an Infantile Neurosis," in *The Standard Edition of the Complete Works of Sigmund Freud,* ed. and trans. James Strachey (London: Hogarth Press, 1953), 3583.

3. Freud, "Three Essays on the History of Sexuality," in ibid., 1485, 1516.

4. See Cătălin Avramescu, *An Intellectual History of Cannibalism,* trans. Alistair I. Blyth (Princeton, NJ: Princeton University Press, 2009).

5. Freud, *The Future of an Illusion,* trans. James Strachey (New York: W.W. Norton, 1961), 10.

6. Ibid., 11.

7. In his late essay "We Are All Cannibals," Claude Lévi-Strauss likewise argued against the notion that cannibalism has been fully banished from civilized life. He distinguished between dramatic cases of "exocannibalism" (often seen in populations who might capture and consume their enemies) and less dramatic cases of "endocannibalism," meaning the ritual preparation and ingestion of parts of the bodies of deceased relatives. Modern societies may not practice ritual endocannibalism, Lévi-Strauss wrote, but modern medicine has begun to reintroduce it by other means. Organ transplants and other techniques that involve the injection or surgical inclusion of somatic tissue from one human body into another, he argued, suggest that endocannibalism is much more widespread than we might think. See Lévi-Strauss, "We Are All Cannibals," in *We Are All Cannibals and Other Essays,* trans. Jane M. Todd (New York: Columbia University Press, 2016). On the anthropological implications of transplants and other medical techniques, see also Sarah Franklin and Margaret Lock (eds.), *Remaking Life & Death: Toward an Anthropology of the Biosciences* (Santa Fe, NM: School of American Research Press, 2003).

8. Freud, *Moses and Monotheism,* trans. Katherine Jones (London: Hogarth Press, 1937), 131–132.

CHAPTER SEVENTEEN. GATHERING/PARTING

1. Claude Lévi-Strauss, *The Elementary Structures of Kinship (Les structures élémentaires de la parenté),* trans. James Harle Bell and John Richard von Sturmer, ed. Rodney Needham (Boston: Beacon Press, 1969).

2. Ibid., 32.

3. The phrasing is Joan Didion's. See *The White Album* (New York: Farrar, Straus and Giroux, 1979), 11.

4. The "carnery" is a very popular image for the future of cultured meat production, promoted by many members of the cultured meat movement, including Isha Datar. See Datar and Robert Bolton, "The Carnery," in *The In Vitro Meat Cookbook* (Amsterdam: Next Nature Network, 2014). The origins of the "carnery" concept are uncertain, but the Reuters journalist Harriet McLeod attributes it to scientist Vladimir Mironov in a 2011 article: "South Carolina Scientist Works to Grow Meat in Lab," *Reuters,* January 30, 2011, www.reuters.com/article/us-food-meat-laboratory-feature/south-carolina-scientist-works-to-grow-meat-in-lab-idUSTRE70T1WZ20110130, accessed September 21, 2018.

5. See Peter Schwartz, *The Art of the Long View* (New York: Penguin Random House, 1991); and Nils Gilman, "The Official Future Is Dead! Long Live the Official Future," www.the-american-interest.com/2017/10/30/official-future-dead-long-live-official-future/, accessed November 19, 2017.

6. See Elizabeth Devitt, "Artificial Chicken Grown from Cells Gets a Taste Test—but Who Will Regulate It?" *Science,* March 15, 2017, www.sciencemag.org/news/2017/03/artificial-chicken-grown-cells-gets-taste-test-who-will-regulate-it, accessed December 6, 2017.

7. See Chase Purdy, "The Idea for Lab-Grown Meat Was Born in a POW Camp," *Quartz,* September 24, 2017, https://qz.com/1077183/the-idea-for-lab-grown-meat-was-born-in-a-prisoner-of-war-camp/, accessed November 27, 2017.

8. Questions have been raised regarding whether the use of a patient's stem cells is a reliable way to get around a patient's immune response under all circumstances, particularly circumstances in which genetic modification of cells is involved. See, for example, Effie Apostolou and Konrad Hochedlinger, "iPS Cells Under Attack," *Nature* 474 (2011): 165–166; and Ryoko Araki et al., "Negligible Immunogenicity of Terminally Differentiated Cells Derived from Induced Pluripotent or Embryonic Stem Cells," *Nature* 494 (2013): 100–104.

9. See, for example, the story of Paolo Macchiarini, as told by John Rasko and Carl Power, "Dr. Con Man: The Rise and Fall of a Celebrity Scientist Who Fooled Almost Everyone," *The Guardian,* September 1, 2017, www.theguardian.com/science/2017/sep/01/paolo-macchiarini-scientist-surgeon-rise-and-fall, accessed December 7, 2017.

10. See, for example, Juliet Eilperin, "Why the Clean Tech Boom Went Bust," *Wired,* January 20, 2012, www.wired.com/2012/01/ff_solyndra/, accessed December 8, 2017.

11. Ryan Fletcher, "All-One Activist: Bruce Friedrich of the Good Food Institute," interview with Bruce Friedrich for Dr. Bronner's, accessed November 28, 2017, at www.drbronner.com.

12. On McDonald's in China, see James L. Watson (ed.), *Golden Arches East: McDonald's in East Asia* (Stanford, CA: Stanford University Press, 1997).

13. On cheapness, see Raj Patel and Jason W. Moore, *A History of the World in Seven Cheap Things: A Guide to Capitalism, Nature, and the Future of the Planet* (Oakland: University of California Press, 2018).

14. On this topic see, for example, Erica Hellerstein and Ken Fine, "A Million Tons of Feces and an Unbearable Stench: Life Near Industrial Pig Farms," *The Guardian,* September 20, 2017, www.theguardian.com/us-news/2017/sep/20/north-carolina-hog-industry-pig-farms, accessed December 10, 2017.

15. Winston Churchill, "Fifty Years Hence," *Popular Mechanics,* March 1932.

16. On the origins of Catts and Zurr's work, see Catts and Zurr, "Semi-living Art," in Eduardo Kac (ed.), *Signs of Life: Bio Art and Beyond* (Cambridge, MA: MIT Press, 2007).

17. Ibid.

18. See Devitt, "Artificial Chicken."

19. For a survey of bio-art around the time of Catts and Zurr's "frog legs" project, see Steve Tomasula, "Genetic Art and the Aesthetics of Biology," *Leonardo* 35 (2002): 137–144.

20. The *Worry Dolls* were the first living-tissue-engineered sculptures displayed, and they originally appeared in Linz, Austria, as part of the Ars Electronica festival. They have subsequently been shown around the globe.

21. See Lynn Margulis, *The Origin of Eukaryotic Cells* (New Haven, CT: Yale University Press, 1970); and Ionat Zurr and Oron Catts, "Are the Semi-living Semi-good or Semi-evil?" *Technoetic Arts* 1 (2003): 49, 51, 54, 59.

22. See the discussion of Margulis on symbiogenesis in Stefan Helmreich, *Alien Ocean: Anthropological Voyages in Microbial Seas* (Berkeley: University of California Press, 2009); see ch. 7.

23. The idea of life as a mystery might fall under the broader subheading of the idea of nature, itself, as a veiled mystery. On this subject, see Pierre Hadot, *The Veil of Isis: An Essay on the History of the Idea of Nature* (Cambridge, MA: Harvard University Press, 2006).

24. Philippa Foot, "Moral Arguments," *Mind* 67 (1958): 502–513. See also G. E. M. Anscombe's "Modern Moral Philosophy," which helps establish the basic stalemate in moral philosophy Foot was trying to go beyond: utilitarian arguments about the consequences of actions on one hand, and deontological arguments that associate goodness with rule following on the other. Anscombe, "Modern Moral Philosophy," *Philosophy* 33 (1958): 1–19.

25. See Philippa Foot and Alan Montefiore, "Goodness and Choice," *Proceedings of the Aristotelian Society, Supplementary Volumes* 35 (1961): 45–80.

26. See Foot, "Does Moral Subjectivism Rest on a Mistake?" *Oxford Journal of Legal Studies* 15 (1995): 1–14.

27. See my discussion of Blumenberg's essay in chapter 11.

28. See Wim Verbeke et al., "Would You Eat 'Cultured Meat?': Consumers' Reactions and Attitude Formation in Belgium, Portugal and the United Kingdom," *Meat Science* 102 (2015): 49–58.

29. Cor van der Weele and Clemens Driessen, "Emerging Profiles for Cultured Meat; Ethics through and as Design," *Animals* 3 (2013): 647–662.

1. The advertisement crossed my Internet browser in the former of a Twitter.com "tweet" from Josh Tetrick, CEO of Just (formerly Hampton Creek), dated October 17, 2018. Tetrick's "400,000" suggests an earlier speciation date for *Homo sapiens* than many experts agree on, but also a late date for the inclusion of meat in the diet of hominins more broadly.

2. Neil Stephens has reminded me that, in one talk he gave on "clean meat," Paul Shapiro (see chapter 15) wielded a harpoon, another case of the ballistic imagination in evidence.

3. See Plato, *Protagoras,* trans. Benjamin Jowett (Indianapolis, IN: Bobbs-Merrill, 1956), lines 320c–328d, pp. 18–19. The list of animal attributes I have included above involves some poetic license; Plato's list of Epimetheus's gifts includes "close hair and thick skins sufficient to defend them against the winter cold and able to resist the summer heat, so that they might have a natural bed of their own when they wanted to rest; also he furnished them with hoofs and hair and hard and callous skins under their feet."

4. Coral Davenport, "Major Climate Report Describes a Strong Risk of Crisis as Early as 2040," *The New York Times,* October 7, 2018. The report Davenport covers here is Myles Allen et al., "Global Warming of 1.5 °C," for the United Nations Intergovernmental Panel on Climate Change, October 6, 2018.

5. See J. Poore and T. Nemecek, "Reducing Food's Environmental Impacts through Producers and Consumers," Science 360 (2018): 987–992.

6. Leo Strauss, *Natural Right and History* (Chicago, IL: University of Chicago Press, 1953), 117.

7. William Gibson, "The Art of Fiction," interview in *The Paris Review,* no. 197, summer 2011.

8. William Gibson to author, exchange via Twitter, 2016.

9. "Speck-ulative fiction," if you're so inclined.

10. See Alvin and Heidi Toffler, *Future Shock* (New York: Random House, 1970).

SELECTED BIBLIOGRAPHY

Adams, Carol. *The Sexual Politics of Meat: A Feminist-Vegetarian Critical Theory* (New York: Continuum, 1990).

Aiello, Leslie C., and Peter Wheeler. "The Expensive-Tissue Hypothesis: The Brain and the Digestive System in Human and Primate Evolution." *Current Anthropology* 36 (1995): 199–221.

Albritton Jonsson, Fredrik. "Island, Nation, Planet: Malthus in the Enlightenment," in Robert Mayhew, ed., *New Perspectives on Malthus* (Oxford, UK: Oxford University Press).

———. "The Origins of Cornucopianism: A Preliminary Genealogy." *Critical Historical Studies* 1 (2014): 151–168.

Albury, W. R. "Politics and Rhetoric in the Sociobiology Debate." *Social Studies of Science* 10 (1980): 519–536.

Alvard, Michael S., and Lawrence Kuznar. "Deferred Harvests: The Transition from Hunting to Animal Husbandry." *American Anthropologist* 103 (2001): 295–311.

Andersson, Jenny. "The Great Future Debate and the Struggle for the World." *The American Historical Review* 117 (2012): 1411–1430.

Anscombe, G. E. M. "Modern Moral Philosophy." *Philosophy* 33 (1958): 1–19.

Arendt, Hannah. *The Human Condition* (Chicago, IL: University of Chicago Press, 1958).

Aristotle. *Politics* (Chicago, IL: University of Chicago Press, 2013).

Arrighi, Giovanni. *The Long Twentieth Century: Money, Power, and the Origins of Our Times* (New York: Verso, 1994).

Avramescu, Cătălin. *An Intellectual History of Cannibalism,* trans. Alistair I. Blyth (Princeton, NJ: Princeton University Press, 2009).

Bachelard, Gaston. *The Psychoanalysis of Fire,* trans. Alan C. M. Ross (London: Routledge & Kegan Paul, 1964).

Bakker, Egbert J. *The Meaning of Meat and the Structure of the Odyssey* (Cambridge, UK: Cambridge University Press, 2013).

Banham, Reyner. *Los Angeles: The Architecture of Four Ecologies* (Harmondsworth, UK: Penguin Books, 1971).

Belasco, Warren. "Algae Burgers for a Hungry World? The Rise and Fall of Chlorella Cuisine." *Technology and Culture* 38 (1997): 608–634.

———. *Meals to Come: A History of the Future of Food* (Berkeley: University of California Press, 2006).

Bell, Daniel. *The Coming of Post-industrial Society: A Venture in Social Forecasting* (New York: Basic Books, 1973).

Bell, Daniel, and Steven Graubard. *Toward The Year 2000: Work in Progress,* special issue of *Daedalus* (1967) [republished by MIT Press in 1969].

Benjamin, Walter. *Illuminations: Essays and Reflections,* trans. Harry Zohn (New York: Harcourt Brace Jovanovich, 1968).

———. *Reflections: Essays, Aphorisms, Autobiographical Writings* (New York: Harcourt Brace Jovanovich, 1978).

Bentham, Jeremy. "A Comment on the Commentaries and a Fragment on Government," ed. J. H. Burns and H. L. A. Hart, in *The Collected Works of Jeremy Bentham* (Oxford, UK: Oxford University Press, 1970).

Berger, John. "Why Look at Animals?" in *About Looking* (New York: Vintage, 1991).

Berman, Marshall. *All That Is Solid Melts into Air: The Experience of Modernity* (New York: Verso, 1982).

Berson, Josh. *The Meat Question: Animals, Humans, and the Deep History of Food* (Cambridge, MA: MIT Press, forthcoming).

Blumenberg, Hans. *Arbeit am Mythos* (Frankfurt am Main: Suhrkamp, 1979) [published in English as *Work on Myth,* trans. Robert M. Wallace (Cambridge, MA: MIT Press, 1985)].

———. *Care Crosses the River,* trans. Paul Fleming (Stanford, CA: Stanford University Press, 2010).

———. "Imitation of Nature: Toward a Prehistory of the Idea of the Creative Being," trans. Ania Wertz. *Qui Parle* 12(1) (2000): 17–54.

Bouk, Dan, and D. Graham Burnett. "Knowledge of Leviathan: Charles W. Morgan Anatomizes His Whale." *Journal of the Early Republic* 28 (2008): 433–466.

Boyd, William. "Making Meat: Science, Technology, and American Poultry Production." *Technology and Culture* 42 (2001): 631–664.

Brand, Stewart. *The Clock of the Long Now: Time and Responsibility* (New York: Basic Books, 1999).

Brock, William H. *Justus von Liebig: The Chemical Gatekeeper* (Cambridge, UK: Cambridge University Press, 1997).

Brown, Nik. "Hope against Hype—Accountability in Biopasts, Presents, and Futures." *Science Studies* 16 (2003): 3–21.

Buck, John Lossing. "Agriculture and the Future of China." *Annals of the American Academy of Political and Social Science,* November 1, 1930.

Budolfson, Mark B. "Is It Wrong to Eat Meat from Factory Farms? If So, Why?" in Ben Bramble and Bob Fischer, eds., *The Moral Complexities of Eating Meat* (Oxford, UK: Oxford University Press, 2015).

Bunn, Henry T. "Meat Made Us Human," in Peter S. Ungar, ed., *Evolution of the Human Diet* (Oxford, UK: Oxford University Press, 2006).

Canguilhem, Georges. "La théorie cellulaire," in *La connaissance de la vie* (Paris: Vrin, 1989) [published in English as "Cell Theory," trans. Stefanos Geroulanos and Daniela Ginsburg, in Georges Canguilhem, *Knowledge of Life,* ed. Paola Marrati and Todd Meyers (New York: Fordham University Press, 2008)].

Caplan, Arthur. *The Sociobiology Debate* (New York: Harper & Row, 1978).

Cartmill, Matt. *A View to a Death in the Morning* (Cambridge, MA: Harvard University Press, 1993).

Catts, Oron, and Ionat Zurr. "Are the Semi-living Semi-good or Semi-evil?" *Technoetic Arts* 1 (2003).

———. "Disembodied Livestock: The Promise of a Semi-living Utopia." *Parallax* 19 (2013): 101–113.

———. "Ingestion/Disembodied Cuisine." *Cabinet,* no. 16: "The Sea" (Winter, 2004/5).

———. "Semi-living Art," in Eduardo Kac, ed., *Signs of Life: Bio Art and Beyond* (Cambridge, MA: MIT Press, 2007).

Cavanaugh, Jillian R. "Making Salami, Producing Bergamo: The Transformation of Value." *Ethnos* 72 (2007): 149–172.

Churchill, Winston. "Fifty Years Hence." *Popular Mechanics,* March 1932.

Cordain, Loren, S. Boyd Eaton, Anthony Sebastian, Neil Mann, Staffan Lindeberg, Bruce A. Watkins, James H. O'Keefe, and Janette Brand-Miller. "Origins and Evolution of the Western Diet: Health Implications for the 21st Century." *American Journal of Clinical Nutrition* 81 (2005): 341–354.

Crary, Jonathan. *24/7* (New York: Verso, 2013).

Crimmins, James. "Bentham and Utilitarianism in the Early Nineteenth Century," in B. Eggleston and D. Miller, eds., *The Cambridge Companion to Utilitarianism* (New York: Cambridge University Press, 2014).

Cronon, William. *Nature's Metropolis: Chicago and the Great West* (New York: W. W. Norton, 1991).

Datar, Isha, and Mirko Betti. "Possibilities for an In Vitro Meat Production System." *Innovative Food Science & Emerging Technologies* 11 (2010): 13–21.

Devine, Philip. "The Moral Basis of Vegetarianism." *Philosophy* 53 (1978): 481–505.

Didion, Joan. *The White Album* (New York: Farrar, Straus and Giroux, 1979).

Dolin, Eric J. *Leviathan: The History of Whaling in America* (New York: W. W. Norton, 2007).

Douglas, Mary. *Purity and Danger* (London: Routledge, 1966).

Driessen, Clemens, and Michiel Korthals. "Pig Towers and In Vitro Meat: Disclosing Moral Worlds by Design." *Social Studies of Science* 42 (2012): 797–820.

Eaton, S. Boyd, and Melvin Konner. "Paleolithic Nutrition: A Consideration of Its Nature and Current Implications." *New England Journal of Medicine* 312 (1985): 283–289.

Eger, Martin. "Hermeneutics and the New Epic of Science," in William Murdo McRae, ed., *The Literature of Science: Perspectives on Popular Science Writing* (Athens: University of Georgia Press, 1993), 186–212.

Ehrlich, Paul, and Anne Ehrlich. *One With Nineveh: Politics, Consumption, and the Human Future* (Washington, DC: Island Press, 2004).

———. *The Population Bomb* (New York: Ballantine, 1968).

Engels, Friedrich. *The Condition of the Working Class in England in 1844*, trans. Florence Kelley Wischnewetzky (London: George Allen & Unwin, 1892).

Fabian, Johannes. *Time and the Other: How Anthropology Makes Its Object* (New York: Columbia University Press, 2014).

Felman, Shoshana. *The Scandal of the Speaking Body: Don Juan with J. L. Austin, or Seduction in Two Languages* (Stanford, CA: Stanford University Press, 2003).

Ferrarin, Alfredo. "Homo Faber, Homo Sapiens, or Homo Politicus? Protagoras and the Myth of Prometheus." *The Review of Metaphysics* 54 (2000): 289–319.

Fiddes, Nick. *Meat: A Natural Symbol* (London: Routledge, 1991).

Foot, Philippa. "Does Moral Subjectivism Rest on a Mistake?" *Oxford Journal of Legal Studies* 15 (1995): 1–14.

———. "Moral Arguments." *Mind* 67 (1958): 502–513.

Foot, Philippa, and Alan Montefiore. "Goodness and Choice." *Proceedings of the Aristotelian Society, Supplementary Volumes* 35 (1961): 45–80.

Fortun, Mike. "For an Ethics of Promising, or: A Few Kind Words about James Watson." *New Genetics and Society* 24 (2005): 157–174.

———. *Promising Genomics: Iceland and deCODE Genetics in a World of Speculation* (Berkeley: University of California Press, 2008).

Foucault, Michel. "Truth and Juridical Forms," in *Power: Essential Works of Foucault, 1954–1984*, ed. Paul Rabinow (New York: The New Press, 2000).

Fox, Michael. " 'Animal Liberation': A Critique." *Ethics* 88 (1978): 106–118.

Francione, Gary L. *Animals as Persons: Essays on the Abolition of Animal Exploitation* (New York: Columbia University Press, 2008).

———. "On Killing Animals." *The Point,* no. 6 (2013).

Francis, Leslie P., and Richard Norman. "Some Animals Are More Equal Than Others." *Philosophy* 53 (1978): 507–527.

Franklin, Sarah. *Dolly Mixtures: The Remaking of Genealogy* (Durham, NC: Duke University Press, 2007).

Franklin, Sarah, and Margaret Lock, eds. *Remaking Life & Death: Toward an Anthropology of the Biosciences* (Santa Fe, NM: School of American Research Press, 2003).

Franklin, Ursula. *The Real World of Technology* (Toronto: Anansi Press, 1999).

Freese, Lee. "The Song of Sociobiology." *Sociological Perspectives* 37 (1994): 337–373.

Freud, Sigmund. *The Future of an Illusion,* trans. James Strachey (New York: W. W. Norton, 1961).

———. *Moses and Monotheism,* trans. Katherine Jones (London: Hogarth Press, 1937).

———. *The Standard Edition of the Complete Works of Sigmund Freud,* ed. and trans. James Strachey (London: Hogarth Press, 1953).

Frey, R. G. *Rights, Killing, and Suffering: Moral Vegetarianism and Applied Ethics* (Oxford, UK: Blackwell, 1983).

Gewertz, Deborah, and Frederick Errington. *Cheap Meat: Flap Food Nations in the Pacific Islands* (Berkeley: University of California Press, 2010).

Gigante, Denise. *Life: Organic Form and Romanticism* (New Haven, CT: Yale University Press, 2009).

Gilman, Nils. *Mandarins of the Future: Modernization Theory in Postwar America* (Baltimore, MD: Johns Hopkins University Press, 2003).

Gordon, Theodore J. "The Methods of Futures Research." *Annals of the American Academy of Political and Social Science* 522 (1992): 25–35.

Gordon, Theodore J., and Olaf Helmer. "Report on a Long-Range Forecasting Study" (Santa Monica, CA: Rand Corporation, 1964).

Hadot, Pierre. *The Veil of Isis: An Essay on the History of the Idea of Nature* (Cambridge, MA: Harvard University Press, 2006).

Hahn Niman, Nicolette. *Defending Beef: The Case for Sustainable Meat Production* (Chelsea, VT: Chelsea Green, 2014).

Hannerz, Ulf. *Writing Future Worlds: An Anthropologist Explores Global Scenarios* (London: Palgrave, 2016).

Haraway, Donna. *Simians, Cyborgs, and Women: The Reinvention of Nature* (New York: Routledge, 1991).

Helmer, Olaf. "Science." *Science Journal* 3(10) (1967): 49–51.

Helmreich, Stefan. *Alien Ocean: Anthropological Voyages in Microbial Seas* (Berkeley: University of California Press, 2009).

———. "Potential Energy and the Body Electric: Cardiac Waves, Brain Waves, and the Making of Quantities into Qualities." *Current Anthropology* 54 (Supplement 7) (2013).

Hesketh, Ian. "The Story of Big History." *History of the Present* 4 (2014): 171–202.

Hopkins, Patrick D., and Austin Dacey. "Vegetarian Meat: Could Technology Save Animals and Satisfy Meat Eaters?" *Journal of Agricultural and Environmental Ethics* 21 (2008): 579–596.

Horowitz, Roger. *Kosher USA: How Coke Became Kosher and Other Tales of Modern Food* (New York: Columbia University Press, 2016).

Horowitz, Roger, Jeffrey M. Pilcher, and Sydney Watts. "Meat for the Multitudes: Market Culture in Paris, New York City, and Mexico City over the Long Nineteenth Century." *American Historical Review* 109 (2004): 1055–1083.

Jameson, Fredric. *Archaeologies of the Future: The Desire Called Utopia and Other Science Fictions* (New York: Verso, 2005).

Jasanoff, Sheila. *Designs on Nature: Science and Democracy in Europe and the United States* (Princeton, NJ: Princeton University Press, 2007).

Johnson, Christopher. "Bricoleur and Bricolage: From Metaphor to Universal Concept." *Paragraph* 35 (2012): 355–372.

Jumonville, Neil. "The Cultural Politics of the Sociobiology Debate." *Journal of the History of Biology* 35 (2002): 569–593.

Kant, Immanuel. *Kant: Political Writings,* ed. H. S. Reiss (Cambridge, UK: Cambridge University Press, 1970).

Kaye, Howard L. *The Social Meaning of Modern Biology: From Social Darwinism to Sociobiology* (New Haven, CT: Yale University Press, 1986).

Kirby, David A. "The Future Is Now: Hollywood Science Consultants, Diegetic Prototypes and the Role of Cinematic Narratives in Generating Real-World Technological Development." *Social Studies of Science* 40 (2010): 41–70.

Korsgaard, Christine M. "Getting Animals in View." *The Point,* no. 6 (2013).

Kovarik, Bill. "Henry Ford, Charles Kettering, and the Fuel of the Future." *Automotive History Review,* no. 32 (Spring 1998): 7–27.

———. "Thar She Blows! The Whale Oil Myth Surfaces Again." TheDailyClimate. Org, March 3, 2014.

Kummer, Corby. *The Pleasures of Slow Food* (San Francisco: Chronicle Books, 2002).

Landecker, Hannah. *Culturing Life: How Cells Became Technologies* (Cambridge, MA: Harvard University Press, 2007).

Laudan, Rachel. *Cuisine and Empire: Cooking in World History* (Berkeley: University of California Press, 2013).

———. "A Plea For Culinary Modernism: Why We Should Love New, Fast, Processed Food." *Gastronomica* 1 (2001): 36–44.

Laughlin, William, Richard B. Lee, and Irven deVore (eds.), with Jill Nash-Mitchell. *Man the Hunter* (Chicago, IL: Aldine-Atherton, 1968).

Leach, Helen M. "Human Domestication Reconsidered." *Current Anthropology* 44 (2003): 349–368.

LeGuin, Ursula K. "A Non-Euclidean View of California as a Cool Place to Be," in *Dancing at the Edge of the World* (London: Gollancz, 1989).

Lévi-Strauss, Claude. *The Elementary Structures of Kinship (Les structures élémentaires de la parenté),* trans. James Harle Bell and John Richard von Sturmer, ed. Rodney Needham (Boston: Beacon Press, 1969).

———. *We Are All Cannibals and Other Essays,* trans. Jane M. Todd (New York: Columbia University Press, 2016).

Lewin, Roger. *Human Evolution: An Illustrated Introduction* (Malden, MA: Blackwell, 2005).

Madrigal, Alexis C. "When Will We Eat Hamburgers Grown in Test-Tubes?" *The Atlantic,* August 6, 2013.

Malthus, Thomas R. *An Essay on the Principle of Population* (London: Penguin, 1985).

Manoury, G. "Sociobiology." *Synthese* 5 (1947): 522–525.

Margulis, Lynn. *The Origin of Eukaryotic Cells* (New Haven, CT: Yale University Press, 1970).

Martin, Michael. "A Moral Critique of Vegetarianism." *Reason Papers,* no. 3 (Fall 1976): 13–43.

Martin, Paul, Nik Brown, and Alison Kraft. "From Bedside to Bench? Communities of Promise, Translational Research and the Making of Blood Stem Cells." *Science as Culture* 17 (2008): 29–41.

Martins, Patrick, with Mike Edison. *The Carnivore Manifesto: Eating Well, Eating Responsibly, and Eating Meat* (New York: Little, Brown, 2014).

Maryanski, Alexandra. "The Pursuit of Human Nature by Sociobiology and by Evolutionary Sociology." *Sociological Perspectives* 37 (1994): 375–389.

Marx, Karl. *The Communist Manifesto: With Related Documents* (Boston: Bedford /St. Martin's, 1999).

Marx, Leo. "Does Improved Technology Mean Progress?" *Technology Review,* January 1987, 33–41, 71.

———. *The Machine in the Garden: Technology and the Pastoral Ideal in America* (Oxford, UK: Oxford University Press, 1964).

Mattick, Carolyn S., Amy E. Landis, Braden R. Allenby, and Nicholas J. Genovese. "Anticipatory Life Cycle Analysis of In Vitro Biomass Cultivation for Cultured Meat Production in the United States." *Environmental Science & Technology* 49 (2015): 11941–11949.

McGee, Harold. *On Food and Cooking: The Science and Lore of the Kitchen* (New York: Scribner, 1984).

McInerney, Jeremy. *The Cattle of the Sun: Cows and Culture in the World of the Ancient Greeks* (Princeton, NJ: Princeton University Press, 2010).

McKenna, Maryn. *Big Chicken: The Improbable Story of How Antibiotics Created Modern Farming and Changed the Way the World Eats* (Washington, DC: National Geographic Books, 2017).

McPhee, John. "A Season on the Chalk," in *Silk Parachute* (New York: Macmillan, 2010).

Mead, Margaret. *The World Ahead: An Anthropologist Contemplates the Future* (New York: Berghahn, 2005).

Messeri, Lisa. *Placing Outer Space: An Earthly Ethnography of Other Worlds* (Chapel Hill, NC: Duke University Press, 2016).

Midgely, Mary. "Sociobiology." *Journal of Medical Ethics* 10 (1984): 158–160.

Miller, Daegan. *This Radical Land: A Natural History of American Dissent* (Chicago, IL: University of Chicago Press, 2018).

Mintz, Sidney. *Sweetness and Power: The Place of Sugar in Modern History* (New York: Viking, 1985).

Mokyr, Joel. *The Gift of Athena: Historical Origins of the Knowledge Economy* (Princeton, NJ: Princeton University Press, 2002).

Nestle, Marion. "Paleolithic Diets: A Skeptical View." *Nutrition Bulletin* 25 (2000): 43–47.

Next Nature Network. *The In Vitro Meat Cookbook* (Amsterdam: Next Nature Network, 2014).

Nietzsche, Friedrich. *On the Genealogy of Morals,* trans. Walter Kaufmann (New York: Vintage Books, 1969).

Oshinsky, David M. *Polio: An American Story* (Oxford, UK: Oxford University Press, 2006).

Ozersky, Josh. *The Hamburger* (New Haven, CT: Yale University Press, 2008).

Patel, Raj, and Jason W. Moore. *A History of the World in Seven Cheap Things: A Guide to Capitalism, Nature, and the Future of the Planet* (Oakland: University of California Press, 2018).

Pauly, Philip J. *Controlling Life: Jacques Loeb and the Engineering Ideal in Biology* (Berkeley: University of California Press, 1987).

Paxson, Heather. *The Life of Cheese: Crafting Food and Value in America* (Berkeley: University of California Press, 2012).

Phillips, Siobhan. "What We Talk about When We Talk about Food." *The Hudson Review* 62 (2009): 189–209.

Plato. *Protagoras,* trans. Benjamin Jowett (Indianapolis, IN: Bobbs-Merrill, 1956).

Pohl, Fredrik, and Cyril M. Kornbluth. *The Space Merchants* (New York: Ballantine, 1953).

Raggio, Olga. "The Myth of Prometheus: Its Survival and Metaphormoses up to the Eighteenth Century." *Journal of the Warburg and Courtauld Institutes* 21 (1958): 44–62.

Ramos-Elorduy, Julieta. "Anthropo-Entomophagy: Cultures, Evolution and Sustainability." *Annual Review of Entomology* 58 (2009): 141–160.

Regan, Tom. *The Case for Animal Rights* (Berkeley: University of California Press, 1983).

———. "Fox's Critique of Animal Liberation." *Ethics* 88 (1978): 126–133.

———. "The Moral Basis of Vegetarianism." *Canadian Journal of Philosophy* 5 (1975): 181–214.

———. "Utilitarianism, Vegetarianism, and Animal Rights." *Philosophy & Public Affairs* 9 (1980): 305–324.

Ritvo, Harriet. *The Animal Estate: The English and Other Creatures in the Victorian Age* (Cambridge, MA: Harvard University Press, 1987).

Robertson, Thomas. *The Malthusian Moment: Global Population Growth and the Birth of American Environmentalism* (New Brunswick, NJ: Rutgers University Press, 2012).

Rodgers, Ben. *Beef and Liberty: Roast Beef, John Bull and the English Nation* (London: Vintage, 2004).

Roe Smith, Merritt. "Technology, Industrialization, and the Idea of Progress in America," in K. B. Byrne, ed., *Responsible Science: The Impact of Technology on Society* (New York: Harper & Row, 1986).

Roe Smith, Merritt, and Leo Marx, eds. *Does Technology Drive History? The Dilemma of Technological Determinism* (Cambridge, MA: MIT Press, 1994).

Rostow, Walt W. *The Stages of Economic Growth: A Non-Communist Manifesto* (Cambridge, UK: Cambridge University Press, 1960).

Russell, Bertrand. "The Harm That Good Men Do." *Harper's,* October 1926.

Sadler, Simon. "The Dome and the Shack: The Dialectics of Hippie Enlightenment," in Iain Boal, Janferie Stone, Michael Watts, and Cal Winslow, eds., *West of Eden: Communes and Utopia in Northern California* (Oakland, CA: PM Press, 2012).

Sahlins, Marshall. "The Original Affluent Society," in *Stone Age Economics* (Chicago, IL: Aldine-Atherton, 1972).

————. *The Use and Abuse of Biology: An Anthropological Critique of Sociobiology* (Ann Arbor: University of Michigan Press, 1976).

Schaler, Jeffrey A., ed. *Peter Singer Under Fire: The Moral Iconoclast Faces His Critics* (Chicago, IL: Open Court, 2009).

Schell, Orville. *Modern Meat: Antibiotics, Hormones, and the Pharmaceutical Farm* (New York: Vintage, 1978).

Schonwald, Josh. *The Taste of Tomorrow: Dispatches from the Future of Food* (New York: HarperCollins, 2012).

Schopenhauer, Arthur. *The World as Will and Representation,* trans. and ed. Judith Norman, Alistair Welchman, and Christopher Janaway (Cambridge, UK: Cambridge University Press, 2010).

Schrempp, Gregory. "Catching Wrangham: On the Mythology and the Science of Fire, Cooking, and Becoming Human." *Journal of Folklore Research* 48 (2011): 109–132.

Schultz, Bart, and Georgios Varouxakis, eds. *Utilitarianism and Empire* (Lanham, MD: Lexington Books, 2005).

Schwartz, Hillel. *The Culture of the Copy: Striking Likenesses, Unreasonable Facsimiles* (revised and updated) (New York: Zone Books, 2014).

Schwartz, Peter. *The Art of the Long View* (New York: Penguin Random House, 1991).

Segerstråle, Ullica. *Defenders of the Truth: The Battle for Science in the Sociobiology Debate and Beyond* (Oxford, UK: Oxford University Press, 2001).

Shapin, Steven. "Invisible Science." *The Hedgehog Review* 18(3) (2016).

Shelley, Mary. *Frankenstein: The Modern Prometheus* (London: Henry Colburn and Richard Bentley, 1831).

Shipman, Pat. "The Animal Connection and Human Evolution." *Current Anthropology* 51 (2010): 519–538.

Singer, Peter. *Animal Liberation* (New York: HarperCollins, 2002).

————. "Ethics and Sociobiology." *Philosophy & Public Affairs* (1982): 40–64.

————. *The Expanding Circle: Ethics, Evolution, and Moral Progress* (Princeton, NJ: Princeton University Press, 2011).

————. "The Fable of the Fox and the Unliberated Animals." *Ethics* 88 (1978): 119–125.

————. "Utilitarianism and Vegetarianism." *Philosophy & Public Affairs* 9 (1980): 325–337.

Smetana, Sergiy, Alexander Mathys, Achim Knoch, and Volker Heinz. "Meat Alternatives: Life Cycle Assessment of Most Known Meat Substitutes." *International Journal of Life Cycle Assessment* 20 (2015): 1254–1267.

Smil, Vaclav. "Eating Meat: Evolution, Patterns, and Consequences." *Population and Development Review* 28 (2002): 599–639.

————. *Feeding the World: A Challenge for the Twenty-First Century* (Cambridge, MA: MIT Press, 2000).

————. "Population Growth and Nitrogen: An Exploration of a Critical Existential Link." *Population and Development Review* 17 (1991): 569–601.

Stanford, Craig B. *The Hunting Ape* (Princeton, NJ: Princeton University Press, 1999).

Steinfeld, Henning, et al. "Livestock's Long Shadow." FAO (2006), www.fao.org /docrep/010/a0701e/a0701e00.HTM.

Stephens, Neil, and Martin Ruivenkamp. "Promise and Ontological Ambiguity in the *In Vitro* Meat Imagescape: From Laboratory Myotubes to the Cultured Burger." *Science as Culture* 25 (2016): 327–355.

Strathern, Marilyn. "Future Kinship and the Study of Culture." *Futures* 27 (1995): 423–435.

———. *Reproducing the Future: Essays on Anthropology, Kinship and the New Reproductive Technologies* (Manchester, UK: Manchester University Press, 1992).

Strauss, Leo. *Natural Right and History* (Chicago, IL: University of Chicago Press, 1953).

Stulp, Gert, Louise Barrett, Felix C. Tropf, and Melinda Mills. "Does Natural Selection Favour Taller Stature among the Tallest People on Earth?" *Proceedings of the Royal Society B* 282 (2015): 20150211.

Taussig, Karen-Sue, Klaus Hoeyer, and Stefan Helmreich. "The Anthropology of Potentiality in Biomedicine." *Current Anthropology* 54 (Supplement 7) (2013).

Tiger, Lionel, and Robin Fox. *Imperial Animal* (New York: Holt, Rinehart and Winston, 1972).

Toffler, Alvin, ed. *The Futurists* (New York: Random House, 1972).

Toffler, Alvin, and Heidi Toffler. *Future Shock* (New York: Random House, 1970).

Tomasula, Steve. "Genetic Art and the Aesthetics of Biology." *Leonardo* 35 (2002): 137–144.

Townsend, Aubrey. "Radical Vegetarians." *Australasian Journal of Philosophy* 57 (1979): 85–93.

Tsing, Anna. "How to Make Resources in order to Destroy Them (and Then Save Them?) on the Salvage Frontier," in Daniel Rosenberg and Susan Harding (eds.), *Histories of the Future* (Chapel Hill, NC: Duke University Press, 2005).

Tuomisto, Hanna L., and M. Joost Teixeira de Mattos. "Environmental Impacts of Cultured Meat Production." *Environmental Science & Technology* 45 (2011): 6117–6123.

Turner, Fred. *From Counterculture to Cyberculture: Stewart Brand, the Whole Earth Network, and the Rise of Digital Utopianism* (Chicago, IL: University of Chicago Press, 2006).

van der Weele, Cor, and Clemens Driessen. "Emerging Profiles for Cultured Meat; Ethics through and as Design." *Animals* 3 (2013): 647–662.

Verbeke, Wim, Afrodita Marcu, Pieter Rutsaert, Rui Gaspar, Beate Seibt, Dave Fletcher, and Julie Barnett. "Would You Eat 'Cultured Meat'?: Consumers' Reactions and Attitude Formation in Belgium, Portugal and the United Kingdom." *Meat Science* 102 (2015): 49–58.

Verbeke, Wim, Pierre Sans, and Ellen J. Van Loo. "Challenges and Prospects for Consumer Acceptance of Cultured Meat." *Journal of Integrative Agriculture* 14 (2015): 285–294.

Walker, Richard A. *The Country in the City: The Greening of the San Francisco Bay Area* (Seattle: University of Washington Press, 2007).

Wallace, David Foster. "Consider the Lobster." *Gourmet,* August 2004, 50–64.

Watson, James L., ed. *Golden Arches East: McDonald's in East Asia* (Stanford, CA: Stanford University Press, 1997).

Watson, James L. "Meat: A Cultural Biography in (South) China," in Jakob A. Klein and Anne Murcott, eds., *Food Consumption in Global Perspective: Essays in the Anthropology of Food in Honour of Jack Goody* (Basingstoke, UK: Palgrave MacMillan, 2014).

Weber, Max. *The Protestant Ethic and the Spirit of Capitalism,* trans. Talcott Parsons (New York: Routledge, 2001).

Weiss, Brad. *Real Pigs: Shifting Values in the Field of Local Pork* (Durham, NC: Duke University Press, 2016).

Wenz, Peter. "Act-Utilitarianism and Animal Liberation." *The Personalist* 60 (1979): 423–428.

Whitman, Charles H. "Old English Mammal Names." *The Journal of English and Germanic Philology* 6 (1907): 649–656.

Williams, Bernard. "A Critique of Utilitarianism," in J.J.C. Smart and Bernard Williams, *Utilitarianism: For and Against* (Cambridge, UK: Cambridge University Press, 1973).

Williams, Raymond. *Keywords: A Vocabulary of Culture and Society* (London: Croom Helm, 1976).

Wilson, E.O. *Sociobiology: The New Synthesis* (Cambridge, MA: Harvard University Press, 1975).

Winner, Langdon. *The Whale and the Reactor: The Search for Limits in the Age of High Technology* (Chicago, IL: University of Chicago Press, 1986).

Wrangham, Richard. *Catching Fire: How Cooking Made Us Human* (New York: Basic Books, 2010).

———. *Demonic Males: Apes and the Origin of Human Violence,* coauthored with Dale Peterson (New York: Houghton Mifflin, 1996).

Zaraska, Marta. *Meathooked: The History and Science of Our 2.5-Million-Year Obsession with Meat* (New York: Basic Books, 2016).

INDEX

CALIFORNIA STUDIES IN FOOD AND CULTURE

Darra Goldstein, Editor